SpringerBriefs in Statistics

More information about this series at http://www.springer.com/series/8921

Chiara Brombin • Luigi Salmaso
Lara Fontanella • Luigi Ippoliti
Caterina Fusilli

Parametric and Nonparametric Inference for Statistical Dynamic Shape Analysis with Applications

Springer

Chiara Brombin
Department of Psychology
Vita-Salute San Raffaele University
Milano, Milano, Italy

Lara Fontanella
Department of Legal and Social Sciences
University of Chieti-Pescara
Pescara, Italy

Caterina Fusilli
Bioinformatics Unit
Casa Sollievo della Sofferenza-Mendel
Rome, Italy

Luigi Salmaso
Department of Management and Engineering
University of Padova
Padova, Italy

Luigi Ippoliti
Department of Economics
University of Chieti-Pescara
Pescara, Italy

ISSN 2191-544X ISSN 2191-5458 (electronic)
SpringerBriefs in Statistics
ISBN 978-3-319-26310-6 ISBN 978-3-319-26311-3 (eBook)
DOI 10.1007/978-3-319-26311-3

Library of Congress Control Number: 2015955726

Springer Cham Heidelberg New York Dordrecht London

Printed on acid-free paper

Springer International Publishing AG Switzerland is part of Springer Science+Business Media (www.springer.com)

Preface

Statistical shape analysis relates to the geometrical study of random objects where location, rotation and scale information can be removed.

The last 20 years have seen a considerable growth in interest in the statistical theory of shape. This has been the result of a synthesis of various disciplines which are interested in measuring, describing and comparing the shapes of objects.

Much work has been done for static or cross-sectional shape analysis, while considerably less research has focused on dynamic or longitudinal shapes. Statistical analysis of dynamic shapes is a problem with significant challenges due to the difficulty in providing the qualitative and quantitative assessment of shape changes over time, across subjects and, eventually, also over groups of subjects.

In this book, we consider specific inferential issues arising from the analysis of dynamic shapes with the attempt to solve the problems at hand using probability models and nonparametric tests. Models are simple to understand and interpret and provide a useful tool to describe the global dynamics of the landmark configurations. However, because of the non-Euclidean nature of shape spaces, distributions in shape spaces are not straightforward to obtain. Here, we consider distributions in the configuration space, with similarity transformations integrated out. This is a simple approach that allows to define models on landmarks themselves giving rise to derived distributions on shapes. The simplest model for a configuration is to assume that the landmarks follow a multivariate Normal distribution about a mean configuration. Various level of generality can also be assumed for the covariance matrix allowing correlations between landmarks and different time points. In this case, it turns out that the distribution which enables inference from configuration onto the shape space is the offset-normal distribution for temporally correlated shapes.

There are also cases of interest in which the use of a model appears problematic and computationally difficult. For example, this is particularly true when the aim of the analysis requires the identification of subsets of landmarks which best describes the dynamics of a whole configuration. A selection of landmarks can enable us to understand and gain information which may not be noticed with a model including all landmarks. To understand whether landmark positions change significantly over time across subjects and over groups of subjects, we make use of the NonParametric Combination (NPC) tests. The NPC methodology, which allows to build powerful tests in a nonparametric framework, does not require strong underlying assumptions as the traditional parametric competitors and allows to work at a local level to highlight specific areas (*domains*) of a configuration in which we may have systematic differences.

The book has a natural split into two parts, with the first three chapters covering material on the offset-normal shape distribution and the remaining chapters covering the theory of NonParametric Combination (NPC) tests. We have attempted to keep each chapter as self-contained as possible, but some dependencies are of course inevitable. The different chapters offer a collection of applications which are bound together by the theme of this book. They refer to the analysis of the FG-NET (Face and Gesture Recognition Research Network) database with facial expressions. For these data set, it may be desirable to provide a description of the dynamics of the expressions, or testing whether there is a difference between the dynamics of two facial expressions or testing which of the landmarks are more informative in explaining the pattern of an expression.

The book is organized as follows. Chapter 1 is the basic introductory chapter for the rest of the book. It introduces the basic notation and commonly used registration approaches of landmark data on a common coordinate system. In Chap. 2 we assume that the shape data are generated from the induced shape distributions of Gaussian configurations in which the similarity transformations are integrated out. For this probability distribution, we discuss the expectation-maximization (EM) algorithm for parameter estimation. This procedure gives essential results for a likelihood-based approach to statistical inference in shape analysis and provides the basis for making inference in a dynamic setting as described in Chap. 3. This latter chapter, in fact, discusses the difficulties of extending results of Chap. 2 in a dynamic framework. Specifically, it describes the offset-normal shape distributions in a dynamic context and introduces the necessary adjustments of the general update rules of the EM algorithm for general spatio-temporal covariance matrices. Also, in order to represent the shape changes in time and classifying dynamic shapes, it provides a discussion of the use of polynomial regression as well as mixture models. In general, it is shown that the EM approach warrants consideration when modelling the dynamics of shapes. However, unless some model simplifications are assumed, the computational burden of the procedure can limit its use in real applications.

In Chap. 4, we introduce the NonParametric Combination (NPC) methodology of a set of dependent partial tests in the specific context of shape analysis. The basic underlying idea of the methodology is that complex multidimensional testing problems may be reduced to a set of simpler subproblems, each provided with a proper permutation solution. These subproblems can be jointly analysed in order to capture the underlying dependency structure, without specifying the nature of dependence relations among variables. NPC tests are distribution-free and, among good general properties, they enjoy the finite-sample consistency property, thus allowing to obtain efficient solutions for multivariate small sample problems, like those encountered in shape analysis applications. Solutions for two independent sample problems is shown, along with suitable combination algorithm, and general framework for dealing with longitudinal repeated-measures designs is examined. Chapter 5 provides examples of applications of the methodology to the FG-NET data: in particular solutions allowing to study differences between dynamics of facial expressions or to identify landmarks that are more involved in the dynamics will be presented. Finally an NPC solution for assessing shape asymmetry in dynamic data is also presented.

Authors are very grateful to Alfred Kume and Fortunato Pesarin for useful discussions and suggestions on the writing of this book.

Milano, Italy — Chiara Brombin
Padova, Italy — Luigi Salmaso
Pescara, Italy — Lara Fontanella
Pescara, Italy — Luigi Ippoliti
Rome, Italy — Caterina Fusilli
July 2015

Contents

Part I
Offset Normal Distribution for Dynamic Shapes

Chapter 1
Basic Concepts and Definitions

Abstract The shape of an object is the geometrical information remaining after the effects of changes in location, scale and orientation have been removed. Information about the objects may come in different forms, for example as a set of landmarks or as a continuous outline. In this chapter we consider landmark based representations of shapes of two-dimensional objects. A common problem here is estimating a mean shape of the group of objects, describing their differences, or assessing the variability within each group.

One way to work with the shapes of different objects is to first register the landmark data on some common coordinate system. Bookstein (Stat Sci 1:181–242, 1986) and Kendall (Bull Lond Math Soc 16:81–121, 1984), each developed coordinate systems for removing the similarity transformations. Alternatively, Procrustes methods (Goodall, J R Stat Soc Ser B 53:285–339, 1991) may be used to remove the similarity transformations. In this chapter we shall discuss these methods by introducing basic concepts and definitions that will be used throughout the book.

Keywords Statistical shape analysis • Landmark coordinates • Registration • Bookstein coordinates • Procrustes analysis

1.1 Landmark Coordinates and the Configuration Space

Shape analysis is considered a cross-disciplinary field characterized by flexible theory and techniques and it has largely been developed through applications in many fields. Relevant references and reviews on the topic include, for example, Goodall (1991), Le and Kendall (1993), Kent (1994, 1995), Dryden and Mardia (1993), Small (1988), Stoyan et al. (1995), Stoyan and Stoyan (1994) and Mardia (1995), Small (1996), Mardia and Dryden (1989), Bookstein (1991), Lele and Richtsmeier (2001), Slice (2005) and Weber and Bookstein (2011).

In many cases of interest the shape features of objects, or images, are frequently explained by the position of a finite collection of points situated in two or three dimensions. Such points are usually called *landmarks*, because, as the name

C. Brombin et al., *Parametric and Nonparametric Inference for Statistical Dynamic Shape Analysis with Applications*, SpringerBriefs in Statistics,
DOI 10.1007/978-3-319-26311-3_1

suggests, they serve as reference points for a partial geometric description of an object. Landmarks are basically classified into the following three groups (Dryden and Mardia 1998):

- *Anatomical landmarks*: these are points assigned by an expert that correspond between organisms in some biologically meaningful way, e.g. the corner of an eye or the meeting of two sutures on a skull.
- *Mathematical landmarks*: these are points located on an object according to some mathematical or geometrical property of the figure, e.g. a high curvature point or an extreme point.
- *Pseudo-landmarks*: these are constructed points on an object, located either around the outline or in between anatomical or mathematical landmarks. Continuous curves can be approximated by a large number of pseudo-landmarks along the curve. Also, pseudo-landmarks are useful in matching surfaces, when points can be located on a regular grid over each surface.

A set of landmarks is said to be labelled if the correspondence of landmarks between different objects is known. That is, a given landmark on one object is known to correspond to a specific landmark on another object. If the correspondences are unknown, the landmarks are said to be unlabelled. Throughout this book we shall work with a configuration of K *labeled landmarks* in a plane.

Definition (Dryden and Mardia 1998). A configuration is a set of landmarks on a particular object. Assuming that the number of not-all-coincident landmarks under study is K, the configuration matrix $\mathbf{X}^\dagger$ is the $K \times m$ matrix of Cartesian coordinates of the K landmarks defined in m dimensions. The **configuration space** is the space of all possible landmark coordinates.

Assuming $m = 2$, the configuration matrix is thus given by a set of Cartesian coordinates $(x_j^\dagger, y_j^\dagger), j = 1, \ldots, K$, leading to a $K \times 2$ matrix

$$\mathbf{X}^\dagger = \begin{pmatrix} x_1^\dagger \; x_2^\dagger \; \ldots \; x_K^\dagger \\ y_1^\dagger \; y_2^\dagger \; \ldots \; y_K^\dagger \end{pmatrix}^T.$$

Note that not only is the case $m = 2$ of particular practical importance, but the fact that we can identify $\mathbb{R}^2$ with $\mathbb{C}$ means that the algebra and geometry of complex numbers can be used to give a neat description of the shape space. Let $z_j^\dagger = x_j^\dagger + iy_j^\dagger$, $j = 1, \ldots, K$ be the j–th landmark expressed in complex coordinates, where $i = \sqrt{-1}$. Then,

$$\mathbf{Z}^\dagger = \boldsymbol{x}^\dagger + i\boldsymbol{y}^\dagger$$

is the $K \times 1$ complex vector of coordinates in the configuration space. Also, since

$$x_j^\dagger = \rho_j \cos\theta_j, \quad y_j^\dagger = \rho_j \sin\theta_j \qquad j = 1, \ldots, k$$

where $\rho_j = \sqrt{\left(x_j^\dagger\right)^2 + \left(y_j^\dagger\right)^2}$ and $\theta_j = \arctan\left(y_j^\dagger / x_j^\dagger\right)$, it follows that

$$z_j^\dagger = \rho_j(\cos\theta_j + i\sin\theta_j) = \rho_j e^{i\theta_j} \qquad j = 1, \ldots, k$$

represents an equivalent representation in polar coordinates. Henceforth, we will thus denote with $\mathbf{X}^\dagger$ the configuration matrix expressed in Cartesian coordinates and with $\mathbf{Z}^\dagger$ the configuration of landmarks expressed in the corresponding complex coordinate system.

1.2 Shape Space

The particular coordinates of the landmark configuration contain some arbitrary and irrelevant information. Working with a raw configuration, in fact, is not convenient since important information can be obscured by differences due to translation, rotation and scaling effects. Hence, this leads us to the following definition.

Definition (Dryden and Mardia 1998). The shape of a configuration matrix $\mathbf{X}^\dagger$ is all the geometrical information about $\mathbf{X}^\dagger$ that is invariant under Euclidean similarity transformations.

The mathematical properties of the shape space for landmark configurations, usually referred to as Kendall's shape space, have been studied intensively. Comprehensive details of the subject are given, for example, in Kendall (1977), Small (1996) and Dryden and Mardia (1998), and we mainly refer to them for known results.

The Euclidean similarity transformation of a configuration matrix $\mathbf{X}^\dagger$ are the set of translated, rotated and isotropically rescaled $\mathbf{X}^\dagger$, i.e.

$$\left\{\varphi \mathbf{X}^\dagger \mathbf{R} + \mathbf{1}_K \boldsymbol{\tau}' : \varphi \in \mathbb{R}^+, \mathbf{R} \in SO\left(m\right), \boldsymbol{\tau} \in \mathbb{R}^m\right\}$$

where φ is a scale parameter, $\mathbf{R}$ is a rotation matrix, $\boldsymbol{\tau}$ is a m-dimensional translation vector and $\mathbf{1}_k$ is a $K \times 1$ vector of ones.

We note that a rotation of a configuration is given by post-multiplication of the matrix $\mathbf{X}^\dagger$ by a rotation matrix $\mathbf{R}$ which satisfies two conditions: $\mathbf{R}^T\mathbf{R} = \mathbf{R}\mathbf{R}^T = \mathbf{I}_m$ and $|\mathbf{R}| = +1$, where $|\cdot|$ denotes the determinant. The set of all $m \times m$ rotation matrices is known as the *special orthogonal group* $SO\left(m\right)$.

Under the action of the Euclidean similarity transformations we can thus define the *shape space*, Ξ_m^K, as the orbit of the non-coincident K point set configurations in $\mathbb{R}^m$ which are invariant under the location, rotation and isotropic scaling (Dryden and Mardia 1998).

Performing similarity transformations leads to losing degrees of freedom and the dimension of the shape space is given by $M = Km - m - 1 - \frac{m(m-1)}{2}$. In fact, for Km

coordinates, we lose m dimensions for location, one dimension for uniform scaling and $\frac{1}{2}m(m-1)$ for rotation; in particular, for $m = 2$, we have $M = 2K - 4$.

For a description of shapes in coordinate terms, consider the $K \times m$ matrix $\mathbf{X}^{\dagger}$. Location and scale effects are easy to eliminate directly.

Translation, for example, can be removed by pre-multiplying the data by the centering matrix

$$\tilde{\mathbf{L}} = \mathbf{I}_K - \mathbf{1}_K \boldsymbol{\delta}^T = \begin{pmatrix} 0 & 0 & 0 & \dots & 0 \\ -1 & 1 & 0 & \dots & 0 \\ -1 & 0 & 1 & \dots & 0 \\ \vdots & \vdots & \vdots & \ddots & \vdots \\ -1 & 0 & 0 & \dots & 1 \end{pmatrix}$$

where $\mathbf{I}_K$ is the $K \times K$ identity matrix, $\mathbf{1}_K$ is a vector of ones of dimension $K \times 1$ and $\boldsymbol{\delta} = (1, 0, \dots, 0)'$ is a $K \times 1$ vector. By removing the first row of $\tilde{\mathbf{L}}$, we define the $(K-1) \times K$ matrix $\mathbf{L}$ and obtain the *pre-form matrix* (i.e. the configuration which is invariant under location shifts of the raw configuration) as $\mathbf{X} = \mathbf{L}\mathbf{X}^{\dagger}$. Thus, standardizing with respect to location allows us to work in a *preform space* of translated configurations denoted by $\mathbf{X}$. Note however that location can also be removed by pre-multiplying by the $(K-1) \times K$ Helmert sub-matrix $\mathbf{H}$ (Dryden and Mardia 1998) or by using the centering matrix

$$\mathbf{C} = \mathbf{I}_K - \frac{1}{K}\mathbf{1}_K \mathbf{1}_K'.$$

The scale effect is related to the concept of size. A *size measure*, $g\left(\mathbf{X}^{\dagger}\right)$, is any positive real valued function of the configuration matrix $\mathbf{X}^{\dagger}$ such that

$$g\left(\varphi \mathbf{X}^{\dagger}\right) = \varphi g\left(\mathbf{X}^{\dagger}\right)$$

for any positive scalar φ. The size measure that is most commonly used in shape analysis is the *centroid size* which is defined as

$$S\left(\mathbf{X}^{\dagger}\right) = \left\|\mathbf{C}\mathbf{X}^{\dagger}\right\|_F = \sqrt{\sum_{j=1}^{k}\sum_{d=1}^{m}\left(\mathbf{x}_{jd}^{\dagger} - \bar{\mathbf{x}}_d^{\dagger}\right)^2}, \qquad \mathbf{X}^{\dagger} \in \mathbb{R}^{Km}$$

where $\mathbf{C}$ is the centering matrix, $\left\|\mathbf{X}^{\dagger}\right\|_F = \sqrt{trace\left(\mathbf{X}^{\dagger T}\mathbf{X}^{\dagger}\right)}$ is the Frobenius norm, $\mathbf{x}_{jd}^{\dagger}$ is the (j, d)th entry of $\mathbf{X}^{\dagger}$ and $\bar{\mathbf{x}}_d^{\dagger}$ is the arithmetic mean in the dth dimension, i.e.

$$\bar{\mathbf{x}}_d^{\dagger} = \frac{1}{k}\sum_{j=1}^{k}\mathbf{x}_{jd}^{\dagger}.$$

Scale effects can thus be removed by dividing by the Euclidean norm of $\mathbf{X}$, i.e. $\mathbf{X}/\|\mathbf{X}\|$, describing a *pre-shape matrix* which defines a point on the unit sphere in $\mathbb{R}^{(k-1)m}$. Hence, one route to obtaining shape coordinates, is first to remove the effect of location and scale, giving *pre-shape*. However, in practice, it is possible to change the order of the filtering operations. In fact, removing only rotation and location involves working with the *size-and-shape* (i.e. *the form*) of $\mathbf{X}^\dagger$. If size is removed from the size-and-shape, then we obtain *the shape* of $\mathbf{X}^\dagger$ which is denoted as

$$[\mathbf{X}] = \left\{\varphi\mathbf{X}\mathbf{R} : \varphi \in \mathbb{R}^+, \mathbf{R} \in SO(m)\right\}.$$

1.3 Coordinate Systems in Two Dimensions

1.3.1 Kendall's Shape Coordinates

To define shape in terms of complex coordinates, let $\mathbf{Z}^\dagger$ denote K complex numbers representing the configuration of raw landmarks in the plane. Then, we can remove the effect of location by moving the sample mean $\bar{z}$ to the origin by a Helmert transformation or, equivalently, by working with the shifted configuration, $\mathbf{Z} = \mathbf{L}\mathbf{Z}^\dagger = (z_2, \ldots, z_K)'$. Next, we observe that the effects of scale and rotation are removed by regarding $(z_2, \ldots, z_K)'$ as equivalent to $c\mathbf{Z} = (cz_2, \ldots, cz_K)'$ for any non-zero complex number c. Thus we can represent the shape of $\mathbf{Z}^\dagger$ by the equivalence class $[z_2, \ldots, z_K]$ of $(z_2, \ldots, z_K)$ under the above equivalence relation. The set of such equivalence classes forms the $(K-2)$–dimensional complex projective space $\mathbb{C}P^{(K-2)}$ with the identification $\Xi_2^K = \mathbb{C}P^{(K-2)}$. Hence, the $(K-2)$ dimensional complex projective space $\mathbb{C}P^{(K-2)}$ can be considered as the unit sphere in $\mathbb{C}^{(K-1)}$, with $\mathbf{Z}$ equivalent to $\mathbf{Z}e^{i\theta}$ for all θ. For example, in the case $K = 3$, the mapping $\mathbf{Z} = (z_2, z_3) \to z_3/z_2$ identifies the complex projective line $\mathbb{C}P^1$ with the sphere S^2 obtained by adding a point at infinity to the complex plane. Thus, we can identify the space Ξ_2^3 of shapes of triangles in the plane with the sphere S^2. Note that, in general, for K points in $\mathbb{R}^2$, the $\boldsymbol{\xi} = z_j/z_2$, $j = 3, \ldots, K$, where z_j are the elements of $\mathbf{Z}$, are referred to as Kendall's shape coordinates. The corresponding $(2K-4) \times 1$ Cartesian coordinates vector of $\boldsymbol{\xi}_u$ is defined as $\mathbf{u} = \left(\Re(\xi_3), \ldots, \Re(\xi_K), \Im(\xi_3), \ldots, \Im(\xi_K)\right)'$.

On the shape space Ξ_2^K several other coordinate systems are in use. In the following, we shall mainly focus on Bookstein and Procrustes coordinates.

1.3.2 Bookstein Coordinates

One popular set of coordinates is represented by Bookstein's shape coordinates (Bookstein 1991). In general, for $m = 2$, translation is achieved by mapping the first point (x_1, y_1) to the origin $(0, 0)$ using the transformation matrix $\mathbf{L}$ which generates the coordinates of the remaining $K - 1$ vertices. This is a linear projection from $\mathbb{R}^{2K}$ to $\mathbb{R}^{2(K-1)}$. The choice of the landmark to be moved to the origin is arbitrary and, in fact, the operation of translation can be done on any other landmark of the configuration.

The *pre-form* configuration of $\mathbf{X}^\dagger$ is given by the $(K - 1) \times 2$ matrix

$$\mathbf{X} = \mathbf{L}\mathbf{X}^\dagger = \begin{pmatrix} -1\ 1\ 0 \dots 0 \\ -1\ 0\ 1 \dots 0 \\ \vdots\ \vdots\ \vdots\ \ddots\ \vdots \\ -1\ 0\ 0 \dots 1 \end{pmatrix} \begin{pmatrix} x_1^\dagger & y_1^\dagger \\ x_2^\dagger & y_2^\dagger \\ \vdots & \vdots \\ x_K^\dagger & y_K^\dagger \end{pmatrix} = \begin{pmatrix} x_2 & y_2 \\ x_3 & y_3 \\ \vdots & \vdots \\ x_K & y_K \end{pmatrix}$$

where the j-th row of $\mathbf{X}$ is written as

$$x_j = x_j^\dagger - x_1^\dagger, \qquad y_j = y_j^\dagger - y_1^\dagger, \qquad j = 2, \dots, K.$$

Elimination of the effects of scaling and rotation can be then achieved through the following operation

$$\mathbf{X} \to \mathbf{U} = \varphi \mathbf{X}\mathbf{R} = \frac{1}{x_2^2 + y_2^2} \begin{pmatrix} x_2 & y_2 \\ x_3 & y_3 \\ \vdots & \vdots \\ x_K & y_K \end{pmatrix} \begin{pmatrix} x_2 & -y_2 \\ y_2 & x_2 \end{pmatrix} = \begin{pmatrix} 1 & 0 \\ u_3 & v_3 \\ \vdots & \vdots \\ u_K & v_K \end{pmatrix}$$

for which $\varphi = 1/|\mathbf{R}|$ and $\mathbf{R} = \begin{pmatrix} x_2 & -y_2 \\ y_2 & x_2 \end{pmatrix}$, with $|\mathbf{R}| = x_2^2 + y_2^2$ and $\mathbf{R}\mathbf{R}' = \mathbf{R}'\mathbf{R} = \varphi^{-1}\mathbf{I}_2$. In general, we do not allow $\mathbf{R}$ to be a reflection. The matrix $\mathbf{U}$ represents the *shape coordinate* matrix $[\mathbf{X}]$ of $\mathbf{X}$, and the elements

$$u_j = \frac{x_j x_2 + y_j y_2}{x_2^2 + y_2^2}, \quad v_j = \frac{-x_j y_2 + y_j x_2}{x_2^2 + y_2^2}, \qquad j = 3, \dots, K$$

often collected in a $(2K - 4) \times 1$ vector $\mathbf{u} = \left(u_3, \dots, u_K, v_3, \dots, v_K\right)'$, are known as *reduced* Bookstein coordinates. Instead, the first two landmarks denote the *base* with respect to which the object is registered.

It is also possible to define the inverse transform from the shape to the preform space. Since $\mathbf{U} = \varphi\mathbf{X}\mathbf{R}$ we have

$$\mathbf{U} \to \mathbf{X} = \varphi^{-1}\mathbf{U}\mathbf{R}^{-1} = \begin{pmatrix} 1 & 0 \\ u_3 & v_3 \\ \vdots & \vdots \\ u_K & v_K \end{pmatrix} \begin{pmatrix} x_2 & y_2 \\ -y_2 & x_2 \end{pmatrix} = \begin{pmatrix} x_2 & y_2 \\ x_3 & y_3 \\ \vdots & \vdots \\ x_K & y_K \end{pmatrix}$$

and therefore,

$$x_j = u_j x_2 - v_j y_2, \quad y_j = u_j y_2 + v_j x_2, \qquad j = 3, \ldots, K.$$

1.3.3 Procrustes Coordinates

For most practical applications, the parameters describing the shapes for a sample of homologous landmark configurations are estimated by a Procrustes superimposition. In this section we thus consider Procrustes shape coordinates to register the data. These coordinates are directly related to Kendall pre-shape coordinates. For convenience, consider two configurations with centred pre-shape coordinates $\check{\boldsymbol{\xi}}^{(1)}$ and $\check{\boldsymbol{\xi}}^{(2)}$, where $\check{\boldsymbol{\xi}}^{(n)} = \mathbf{Z}^{(n)} / \left\|\mathbf{Z}^{(n)}\right\|$, $n = 1, 2$. Following Dryden and Mardia (1998), we can match $\check{\boldsymbol{\xi}}^{(1)}$ to $\check{\boldsymbol{\xi}}^{(2)}$ using complex linear regression, minimising the difference between the two pre-shapes using Procrustes analysis arguments.

To measure distances between shapes we require a metric on Ξ_2^K; this gives Kendall's shape space the structure of a Riemannian manifold with Procrustes distance as its metric (Dryden and Mardia 1998). The Procrustes distance, denoted with $\rho\left(\mathbf{Z}^{(1)}, \mathbf{Z}^{(2)}\right)$, is the closest great circle distance between $\check{\boldsymbol{\xi}}^{(1)}$ and $\check{\boldsymbol{\xi}}^{(2)}$ on the pre-shape sphere. Because shapes correspond to points on a hemisphere with unit radius, ρ is also the angle, in radians, between vectors from the center of the hemisphere to the two points being compared.

Alternative distances in shape space can also be used in principle (see Dryden and Mardia 1998, Chap. 3). For example, the *partial Procrustes distance*, $d_P\left(\mathbf{Z}^{(1)}, \mathbf{Z}^{(2)}\right)$, can be obtained by matching the pre-shapes $\check{\boldsymbol{\xi}}^{(1)}$ and $\check{\boldsymbol{\xi}}^{(2)}$ of $\mathbf{Z}^{(1)}$ and $\mathbf{Z}^{(2)}$ as closely as possible over rotations, but not scale. The partial Procrustes distance then can be regarded as the chordal distance between the complex pre-shapes $\check{\boldsymbol{\xi}}^{(1)}$ and $\check{\boldsymbol{\xi}}^{(2)}$. On the other hand, if the matching of the pre-shapes is obtained by minimizing the distance between $\check{\boldsymbol{\xi}}^{(1)}$ and $\check{\boldsymbol{\xi}}^{(2)}$ over scale and rotation, then we have the *full Procrustes distance*, $d_F\left(\mathbf{Z}^{(1)}, \mathbf{Z}^{(2)}\right)$, which is the closest Euclidean distance between $\mathbf{Z}^{(1)}$ and $\mathbf{Z}^{(2)}$.

Figure 1.1 shows a cross section of the pre-shape sphere illustrating the relationship between d_F, d_P and ρ (Dryden and Mardia 1998). Indeed we have:

$$d_F\left(\mathbf{Z}^{(1)}, \mathbf{Z}^{(2)}\right) = \sin\rho \quad \text{and} \quad d_P\left(\mathbf{Z}^{(1)}, \mathbf{Z}^{(2)}\right) = 2\sin\left(\rho/2\right).$$

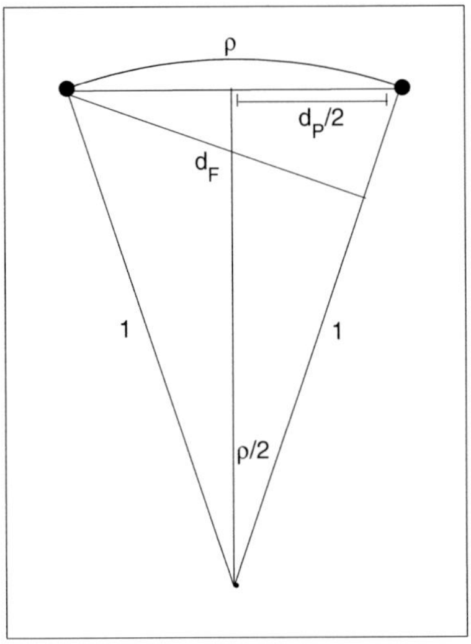

Fig. 1.1 Section of pre-shape sphere, illustrating the relationship between the Procrustes distances d_F, d_P and ρ

The registration procedure can be extended to a set of N configurations in order to find the Procrustes mean shape, $\boldsymbol{\mu}$, and in turn the Procrustes shape coordinates. Let us assume that the configurations $\mathbf{Z}^{\dagger^{(1)}}, \mathbf{Z}^{\dagger^{(2)}}, \ldots, \mathbf{Z}^{\dagger^{(N)}}$ have been centered, so that $\mathbf{Z}^{(1)}, \ldots, \mathbf{Z}^{(N)}$ is the set of configurations in the preform space. Estimation of the mean shape is performed through the Generalized Procrustes analysis (GPA, Dryden and Mardia 1998, p. 88). It can be shown that by using the full Procrustes distances, the estimated mean shape, $\hat{\boldsymbol{\mu}}$, is obtained by minimizing (over $\boldsymbol{\mu}$) the sum of square full Procrustes distances from each $\mathbf{Z}^{(n)}$ to an unknown unit size mean configuration $\boldsymbol{\mu}$ ($\|\boldsymbol{\mu}\| = 1$), i.e.

$$\hat{\boldsymbol{\mu}} = \arg\inf_{\boldsymbol{\mu}} \sum_{n=1}^{N} d_F^2 \left(\mathbf{Z}^{(n)}, \boldsymbol{\mu}\right).$$

For two-dimensional data (i.e. $m = 2$), an explicit eigenvector solution exists for $\hat{\boldsymbol{\mu}}$, since it can be found as the eigenvector corresponding to the largest eigenvalue of the complex sum of squares and products matrix (Kent 1994)

$$\mathbf{S} = \sum_{l=1}^{n} \check{\boldsymbol{\xi}}^{(n)} \check{\boldsymbol{\xi}}^{*(n)}$$

where $\check{\boldsymbol{\xi}}^{*(n)}$ is the transpose of the complex conjugate of $\check{\boldsymbol{\xi}}^{(n)}$. The eigenvector is unique (up to a rotation) provided there is a single largest eigenvalue of $\mathbf{S}$. For higher dimensions an iterative procedure is required (see Dryden and Mardia 1998, p. 90).

Because all differences in location, scale, and orientation have been removed by this procedure, any differences in coordinates of corresponding landmarks between configurations must be the result of differences in shape between those configurations. The *full Procrustes coordinates* of $\mathbf{Z}^{(1)}, \ldots, \mathbf{Z}^{(N)}$ are then defined as

$$\boldsymbol{\xi}_P^{(n)} = \frac{\mathbf{Z}^{*(n)}\hat{\boldsymbol{\mu}}\mathbf{Z}^{(n)}}{\mathbf{Z}^{*(n)}\mathbf{Z}^{(n)}} = \check{\boldsymbol{\xi}}^{*(n)}\hat{\boldsymbol{\mu}}\check{\boldsymbol{\xi}}^{(n)} \qquad n = 1, \ldots, N$$

where $\boldsymbol{\xi}_P^{(n)}$ is the full Procrustes fit of $\mathbf{Z}^{(n)}$ onto $\hat{\boldsymbol{\mu}}$. Note that the full Procrustes mean shape can also be obtained as an arithmetic mean of the full Procrustes coordinates. Hence, the *Procrustes residuals*, which are useful for investigating shape variability, can be obtained as

$$\boldsymbol{\xi}_R^{(n)} = \boldsymbol{\xi}_P^{(n)} - \left(\frac{1}{N}\sum_{n=1}^{N}\boldsymbol{\xi}_P^{(n)}\right) \qquad n = 1, \ldots, N. \tag{1.1}$$

Procrustes superimposition places configurations in a non-Euclidean shape space and this makes statistical inference more difficult than in Euclidean spaces. Because most statistical methods are predicated on Euclidean relationships between variables, most analyses of shape data involve projecting data from this space into a Euclidean space tangent to the shape space at a pole $\boldsymbol{\gamma}$. We recall that Kendall pre-shape space is a unit complex sphere. Then, the tangent plane to this sphere at $\boldsymbol{\gamma}$ is a linearised version of the shape space. The pole $\boldsymbol{\gamma}$ usually corresponds to the average shape, for example the Procrustes mean shape. Rotating and scaling the pre-shapes to minimise the Euclidean distance to the pole, and then projecting on to the tangent plane, gives Procrustes tangent coordinates. When variation in shape is sufficiently small, it is thus possible to replace Kendall's shape space with a Euclidean approximation (Kent 1994).

Specifically, we can define departures of each data shape from $\hat{\boldsymbol{\mu}}$ in terms of *full Procrustes tangent coordinates* as (Dryden and Mardia 1998)

$$\boldsymbol{\xi}_v^{(n)} = \hat{\varphi}e^{-i\hat{\theta}}\left[\mathbf{I}_{K-1} - \hat{\boldsymbol{\mu}}\hat{\boldsymbol{\mu}}^*\right]\check{\boldsymbol{\xi}}^{(n)}, \quad \boldsymbol{\xi}_v^{(n)} \in T(\hat{\boldsymbol{\mu}}), \; n = 1, \ldots, N$$

where $\hat{\varphi} = \|\hat{\boldsymbol{\mu}}^*\check{\boldsymbol{\xi}}^{(n)}\|$, $\hat{\theta} = Arg(\hat{\boldsymbol{\mu}}^*\check{\boldsymbol{\xi}}^{(n)})$ and $T(\hat{\boldsymbol{\mu}})$ is the plane tangent at the pole $\hat{\boldsymbol{\mu}}$. Each tangent vector $\boldsymbol{\xi}_v^{(n)}$ lies in $\mathbb{C}^K$ and hence has $2K$ components. But due to several linear constraints (i.e. $\boldsymbol{\xi}_v^{*(n)}\mathbf{1} = 0$ prohibiting translation and $\boldsymbol{\xi}_v^{*(n)}\hat{\boldsymbol{\mu}} = 0$ prohibiting rotation and scaling) which prohibit directions of change which would only affect the registration of $\check{\boldsymbol{\xi}}^{(n)}$ but not its shape, there are only $2K - 4$ degrees of freedom in $\boldsymbol{\xi}_v^{(n)}$, just as for Bookstein coordinates. Also, note that the Procrustes residuals given in Eq. (1.1) are approximate tangent coordinates. Hence, if a shape is close to the pole, then the differences between the choice of partial, full or Procrustes tangent coordinates or Procrustes residuals will be very small.

In general, deciding how constructing the projection of shapes onto the tangent plane includes deciding: (a) the space representing the source of the configurations projected onto the tangent plane, (b) the rule to be used to determine the direction of

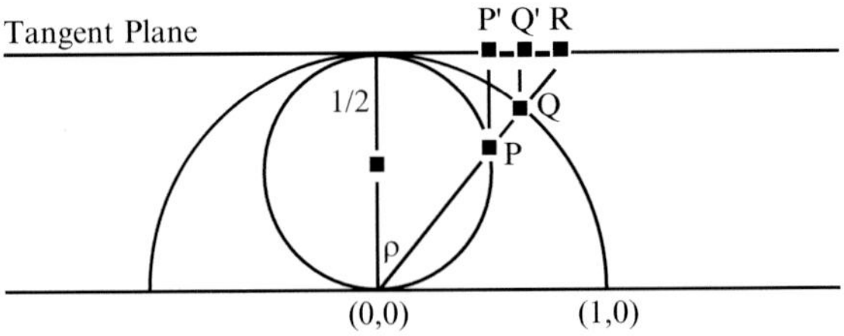

Fig. 1.2 Stereographic and orthogonal projections onto the tangent plane. Point P represents a *triangle* in Kendall's shape space and Q is the same shape scaled to unit centroid size. Point R is a stereographic projection of P and Q onto the tangent plane. Q' and P' are the orthogonal projections of Q and P, respectively, onto the tangent plane

the projection and (c) an appropriate reference (pole) configuration to serve as the tangent point.

The problem of replacing a curved space with a Euclidean approximation can be illustrated for the special case of triangles. The outer hemisphere in Fig. 1.2 is the space constructed by aligning pre-shapes (with centroid size fixed at one) to minimize the partial Procrustes distance. The inner sphere is Kendall's shape space, constructed by scaling the aligned target shapes to centroid size equal to $\cos(\rho)$. These two spaces share a common point, the reference shape $\boldsymbol{\mu}$, because the distance of the reference from itself is zero, so $\cos(\rho)$ is one. Tangent to both of these spaces, at the reference shape (i.e. the *pole*), is a Euclidean plane. One approach to project to the new space is to choose the hemisphere of aligned pre-shapes as the reference space so that the projections are along the radii of this hemisphere to the tangent space. In this stereographic projection, the shape represented by points P and Q (with centroid sizes equal to $\cos(\rho)$ and one, respectively) map to the same location (R) in the tangent space. The distance in the plane from the reference to R is greater than the arc-length from the reference to P (the Procrustes distance); and the discrepancy between these distances increases as ρ increases and the distance in the tangent plane approaches infinity.

The other approach to projecting from one space onto another is to project along lines that are orthogonal to the new space. Point P' represents the orthogonal projection of P onto the tangent plane, and this projection produces distances from the reference in the tangent plane that are less than the Procrustes distance. As in the stereographic projection, the magnitude of the discrepancy between the distances increases as ρ increases, but in the orthogonal projections, distances in the tangent plane asymptotically approach the maximum equal to the radius of the shape space. In the stereographic projection it does not matter whether the projection to the tangent plane is from the hemisphere of triangles in partial Procrustes superimposition, or from the sphere of triangles in full Procrustes superimposition. Both target configurations project to the same point in the tangent space. In the orthogonal projection, it does matter whether the projection to the tangent plane is from the outer or inner hemisphere. The projection from the hemisphere produces distances in the tangent plane that depart less from the Procrustes distance (the arc length) and are closer to the partial Procrustes distance (the chord length). Projection from the sphere produces distances that depart more from the Procrustes distance and are closer to the full Procrustes distance. Furthermore, the projections from the hemisphere of triangles in partial Procrustes superimposition have a higher maximum distance from the reference (one instead of one-half), and approach it more slowly. It can be shown that the differences between the Procrustes, partial Procrustes and full Procrustes distances from the reference become negligible as ρ approaches zero. Similarly, the differences among the stereographic and orthogonal projections also become negligible as ρ approaches zero.

References

Bookstein FL (1986) Size and shape spaces for landmark data in two dimensions. Stat Sci 1:181–242

Bookstein FL (1991) Morphometric tools for landmark data: geometry and biology. Cambridge University Press, Cambridge

Dryden IL, Mardia KV (1993) Multivariate shape analysis. Sankhyā Ser A 55:460–480

Dryden IL, Mardia KV (1998) Statistical shape analysis. Wiley, London

Goodall CR (1991) Procrustes methods in the statistical analysis of shape. J R Stat Soc Ser B 53:285–339

Kendall DG (1977) The diffusion of shape. Adv Appl Probab 9:428–430

Kendall DG (1984) Shape manifolds, procrustean metrics, and complex projective spaces. Bull Lond Math Soc 16:81–121

Kent JT (1994) The complex bingham distribution and shape analysis. J R Stat Soc Ser B 56:285–299

Kent JT (1995) Current issues for statistical inference in shape analysis. In: Mardia KV, Gill CA (eds) Proceedings in current issues in statistical shape analysis. University of Leeds Press, Leeds pp 167–175

Le H, Kendall DG (1993) The riemannian structure of euclidean shape spaces: a novel environment for statistics. Ann Stat 21:1225–1271

Lele SR, Richtsmeier JT (2001) An invariant approach to statistical analysis of shapes. Chapman & Hall/CRC, Boca Raton, FL

Mardia KV (1995) Shape advances and future perspectives. In: Mardia KV, Gill CA (eds) Proceedings in current issues in statistical shape analysis. University of Leeds Press, Leeds, pp 57–75

Mardia KV, Dryden IL (1989) The statistical analysis of shape data. Biometrika 76:271–281

Slice DE (2005) Modern morphometrics in physical anthropology. Springer, New York

Small CG (1988) Techniques of shape analysis on sets of points. Int Stat Rev 56:243–257

Small CG (1996) The statistical theory of shape. Springer, New York

Stoyan D, Stoyan H (1994) Fractals, random shapes and point fields: methods of geometric statistics. Wiley, Chichester

Stoyan D, Kendall WS, Mecke J (1995) Stochastic geometry and its applications, 2nd edn. Wiley, Chichester, England

Weber GW, Bookstein FL (2011) Virtual anthropology: a guide to a new interdisciplinary field. Springer, New York

Chapter 2
Shape Inference and the Offset-Normal Distribution

Abstract In this chapter we work directly with the offset-normal shape distribution as a probability model for statistical inference on a sample of landmark configurations. This enables inference for induced Gaussian processes from configurations onto the shape space. Following Kume and Welling (J Comput Graph Stat 19:702–723, 2010), an Expectation Maximization (EM) algorithm for computing exact maximum likelihood (ML) estimation of the involved parameters is discussed. The chapter concludes with an application on facial expression analysis.

Keywords Offset-normal shape distribution • Shape analysis • EM Algorithm • Facial expression analysis

2.1 Introduction

As shown in the previous chapter, in landmark based shape analysis, coordinate information can be represented by a $K \times m$ configuration matrix, $\mathbf{X}^{\dagger}$, where $m = 2$ in the planar case.

In statistical shape analysis, it is of interest the study of an *iid* sample of planar configurations, $\mathbf{X}^{\dagger^{(1)}}, \ldots, \mathbf{X}^{\dagger^{(N)}}$, generated by some distribution $F(\mathbf{X}^{\dagger})$ and observed after each one of these is randomly rescaled, rotated, and translated. In other words, our observed data consist of elements $\left\{\varphi_n \left(\mathbf{X}^{\dagger^{(n)}} + \mathbf{1}_K \otimes \boldsymbol{\tau}_n'\right) \mathbf{R}_n\right\}_{n=1:N}$, where $\varphi_n > 0$ is a rescaling factor, $\mathbf{R}_n$ is an element from $SO(2)$, the group of rotations in the plane, and $\mathbf{1}_K \otimes \boldsymbol{\tau}_n'$, with $\mathbf{1}_K$ a K-vector of ones and $\otimes$ the Kronecker product, represents the translation effect by a vector $\boldsymbol{\tau}_n$ in the plane. Since, the geometric characteristics of the landmark configuration are of primary interest and the particular coordinate system contains some arbitrary and irrelevant information, φ_n, $\boldsymbol{\tau}_n$, and $\mathbf{R}_n$ can be thought of as nuisance parameters. Therefore, statistical inference based on the underlying distribution $F(\mathbf{X}^{\dagger})$ needs to be invariant to location, rotation, and scaling for each observed element, $\varphi_n \left(\mathbf{X}^{\dagger^{(n)}} + \mathbf{1}_K \otimes \boldsymbol{\tau}_n'\right) \mathbf{R}_n$. This is essentially an inference problem based on the shapes of planar configurations $\mathbf{X}^{\dagger(n)}$.

C. Brombin et al., *Parametric and Nonparametric Inference for Statistical Dynamic Shape Analysis with Applications*, SpringerBriefs in Statistics,
DOI 10.1007/978-3-319-26311-3_2

In this chapter we assume that the probability model is defined by following a marginal approach (Dryden and Mardia 1998, p.109), in that we assume the shape data are generated from the induced shape distributions of Gaussian configurations $\mathbf{X}^\dagger$ in which the similarity transformations are integrated out. These models for two-dimensional shape objects have been originally introduced by Dryden and Mardia (1991) where the shape distribution of $\mathbf{X}^\dagger$ is obtained in closed form. The estimation procedure for the model parameters is based on the Expectation Maximization (EM) algorithm introduced by Kume and Welling (2010) and, for convenience, we mainly refer to them for known results. This procedure gives essential results for a likelihood-based approach for statistical inference in shape analysis and provides the basis for making inference in a dynamic setting. Also, the algorithm is baseline invariant and can be adjusted for missing data (Kume and Welling 2010).

This chapter is organized as follows. Section 2.2 provides a discussion of the Gaussian distribution in the configuration space by considering different covariance structures for the landmarks. Sections 2.3 and 2.4 lead to the specification of the offset-normal shape distribution. In Sect. 2.5 we introduce the EM algorithm establishing the relevant update rules for the model parameters. Section 2.6 finally concludes the chapter with an application on the FG-NET (Face and Gesture Recognition Research Network) database with facial expressions and emotions.

2.2 The Gaussian Distribution in the Configuration Space

Given the configuration $\mathbf{X}^\dagger \in \Re^{K\times 2}$, we assume that the vector obtained by stacking the columns of the configuration matrix on top of one another follows a general $2k$-dimensional Gaussian distribution with a $2K$-dimensional mean vector, $\boldsymbol{\mu}^\dagger$, and a $2K \times 2K$ covariance matrix, $\boldsymbol{\Sigma}^\dagger$, that is

$$vec(\mathbf{X}^\dagger) = \begin{pmatrix} \mathbf{x}^\dagger \\ \mathbf{y}^\dagger \end{pmatrix} \sim \mathcal{N}_{2K}\left[vec(\boldsymbol{\mu}^\dagger) = \begin{pmatrix} \boldsymbol{\mu}_x^\dagger \\ \boldsymbol{\mu}_y^\dagger \end{pmatrix}, \boldsymbol{\Sigma}^\dagger = \begin{pmatrix} \boldsymbol{\Sigma}_{xx}^\dagger & \boldsymbol{\Sigma}_{xy}^\dagger \\ \boldsymbol{\Sigma}_{yx}^\dagger & \boldsymbol{\Sigma}_{yy}^\dagger \end{pmatrix}\right]$$

where $vec(\cdot)$ is the *vec* operator and $\boldsymbol{\Sigma}_{yx}^\dagger = \boldsymbol{\Sigma}_{xy}^{\dagger\prime}$.

In statistical shape analysis, it is common to consider covariance structures which are invariant with respect to rotation. The more general form of rotational invariant covariance structure is represented by the complex covariance. Let $\mathbf{z}^\dagger = \mathbf{x}^\dagger + i\mathbf{y}^\dagger \in \mathbb{C}^K$ be the complex vector in the configuration space. In order to express in complex notation the covariance matrix, $\boldsymbol{\Sigma}^\dagger$, the following two different matrices have to be considered:

- the covariance matrix

$$\boldsymbol{\Sigma}_z^\dagger = E[(\mathbf{z}^\dagger - \boldsymbol{\mu}_z^\dagger)(\mathbf{z}^\dagger - \boldsymbol{\mu}_z^\dagger)^*] = (\boldsymbol{\Sigma}_{xx}^\dagger + \boldsymbol{\Sigma}_{yy}^\dagger) + i(\boldsymbol{\Sigma}_{xy}^\dagger - \boldsymbol{\Sigma}_{yx}^\dagger)$$

  where $*$ represents the complex conjugate transpose;

- and the pseudo-covariance (Neeser and Massey 1993) matrix,

$$\tilde{\boldsymbol{\Sigma}}_z^{\dagger} = E[(\mathbf{z}^{\dagger} - \boldsymbol{\mu}_z^{\dagger})(\mathbf{z}^{\dagger} - \boldsymbol{\mu}_z^{\dagger})'] = (\boldsymbol{\Sigma}_{xx}^{\dagger} - \boldsymbol{\Sigma}_{yy}^{\dagger}) + i(\boldsymbol{\Sigma}_{xy}^{\dagger} + \boldsymbol{\Sigma}_{yx}^{\dagger}).$$

This matrix is also known as the *relation* (Picinbono 1996) or the *complementary* covariance matrix (Schreier and Scharf 2003; Adali et al. 2011).

To better understand the second order properties of a complex random vector, consider first the vector $vec(\mathbf{X}^{\dagger} - \boldsymbol{\mu}^{\dagger}) = \left(\mathbf{x}_c^{\dagger'}\ \mathbf{y}_c^{\dagger'}\right)'$, where $\mathbf{x}_c^{\dagger} = \mathbf{x}^{\dagger} - \boldsymbol{\mu}_x^{\dagger}$ and $\mathbf{y}_c^{\dagger} = \mathbf{y}^{\dagger} - \boldsymbol{\mu}_y^{\dagger}$ are zero-mean real random vectors; therefore $\mathbf{z}_c^{\dagger} = \mathbf{x}_c^{\dagger} + i\mathbf{y}_c^{\dagger}$ is a zero mean complex random vector.

Then, consider the $2K$-dimensional augmented complex vector ς from $\mathbf{z}_c^{\dagger}$, which is obtained through the transformation (Adali et al. 2011)

$$\varsigma = \begin{pmatrix} \mathbf{z}_c^{\dagger} \\ \bar{\mathbf{z}}_c^{\dagger} \end{pmatrix} = \mathbf{T} \begin{pmatrix} \mathbf{x}_c^{\dagger} \\ \mathbf{y}_c^{\dagger} \end{pmatrix} = \frac{1}{\sqrt{2}} \begin{pmatrix} \mathbf{x}_c^{\dagger} + i\mathbf{y}_c^{\dagger} \\ \mathbf{x}_c^{\dagger} - i\mathbf{y}_c^{\dagger} \end{pmatrix}$$

where $\bar{\mathbf{z}}_c^{\dagger}$ represents the complex conjugate of $\mathbf{z}_c^{\dagger}$, and the unitary matrix $\mathbf{T}$ is given by

$$\mathbf{T} = \frac{1}{\sqrt{2}} \begin{pmatrix} \mathbf{I}_K & +i\mathbf{I}_K \\ \mathbf{I}_K & -i\mathbf{I}_K \end{pmatrix}.$$

The $2K \times 2K$ covariance matrix of the augmented vector ς turns out to be

$$\boldsymbol{\Sigma}_{\zeta} = E[\varsigma\varsigma^*] = E\left[\begin{pmatrix} \mathbf{z}_c^{\dagger} \\ \bar{\mathbf{z}}_c^{\dagger} \end{pmatrix} \begin{pmatrix} \mathbf{z}_c^{\dagger *} & \mathbf{z}_c^{\dagger'} \end{pmatrix}\right] = \begin{pmatrix} \boldsymbol{\Sigma}_z^{\dagger} & \tilde{\boldsymbol{\Sigma}}_z^{\dagger} \\ \tilde{\boldsymbol{\Sigma}}_z^{\dagger *} & \boldsymbol{\Sigma}_z^{\dagger *} \end{pmatrix}.$$

Besides $\boldsymbol{\Sigma}_z^{\dagger}$ being positive definite and $\tilde{\boldsymbol{\Sigma}}_z^{\dagger}$ being symmetric, the Schur complement $\boldsymbol{\Sigma}_z^{\dagger *} - \tilde{\boldsymbol{\Sigma}}_z^{\dagger *} \boldsymbol{\Sigma}_z^{\dagger -1} \tilde{\boldsymbol{\Sigma}}_z^{\dagger}$ must be positive definite to ensure that $\boldsymbol{\Sigma}_{\zeta}$ is positive definite and, thus, a valid covariance matrix for ς.

We can derive an equivalent formulation of $\boldsymbol{\Sigma}_{\zeta}$ by applying the unitary transformation $\mathbf{T}$ onto the covariance matrix in the configuration space, i.e. $\boldsymbol{\Sigma}_{\zeta} = \mathbf{T}\boldsymbol{\Sigma}^{\dagger}\mathbf{T}^*$, which gives

$$\begin{aligned} \boldsymbol{\Sigma}_{\zeta} &= \frac{1}{2} \begin{pmatrix} \mathbf{I}_K & +i\mathbf{I}_K \\ \mathbf{I}_K & -i\mathbf{I}_K \end{pmatrix} \begin{pmatrix} \boldsymbol{\Sigma}_{xx}^{\dagger} & \boldsymbol{\Sigma}_{xy}^{\dagger} \\ \boldsymbol{\Sigma}_{yx}^{\dagger} & \boldsymbol{\Sigma}_{yy}^{\dagger} \end{pmatrix} \begin{pmatrix} \mathbf{I}_K & \mathbf{I}_K \\ +i\mathbf{I}_K & -i\mathbf{I}_K \end{pmatrix} \\ &= \frac{1}{2} \begin{pmatrix} \boldsymbol{\Sigma}_{xx}^{\dagger} + \boldsymbol{\Sigma}_{yy}^{\dagger} - i(\boldsymbol{\Sigma}_{yx}^{\dagger} - \boldsymbol{\Sigma}_{xy}^{\dagger}) & \boldsymbol{\Sigma}_{xx}^{\dagger} - \boldsymbol{\Sigma}_{yy}^{\dagger} + i(\boldsymbol{\Sigma}_{yx}^{\dagger} + \boldsymbol{\Sigma}_{xy}^{\dagger}) \\ \boldsymbol{\Sigma}_{xx}^{\dagger} - \boldsymbol{\Sigma}_{yy}^{\dagger} - i(\boldsymbol{\Sigma}_{yx}^{\dagger} + \boldsymbol{\Sigma}_{xy}^{\dagger}) & \boldsymbol{\Sigma}_{xx}^{\dagger} + \boldsymbol{\Sigma}_{yy}^{\dagger} - i(\boldsymbol{\Sigma}_{yx}^{\dagger} - \boldsymbol{\Sigma}_{xy}^{\dagger}) \end{pmatrix}. \end{aligned}$$

Therefore, the $K \times K$ covariance matrix $\boldsymbol{\Sigma}_z^\dagger = E\left[\mathbf{z}_c^\dagger \mathbf{z}_c^{\dagger *}\right] = \left(\boldsymbol{\Sigma}_{xx}^\dagger + \boldsymbol{\Sigma}_{yy}^\dagger + i(\boldsymbol{\Sigma}_{yx}^\dagger - \boldsymbol{\Sigma}_{xy}^\dagger)\right)$ alone is not sufficient to describe the second-order behavior of ς. We also need the $K \times K$ symmetric complementary covariance matrix $\tilde{\boldsymbol{\Sigma}}_z^\dagger = E\left[\mathbf{z}_c^\dagger \mathbf{z}_c^{\dagger'}\right] = \left(\boldsymbol{\Sigma}_{xx}^\dagger - \boldsymbol{\Sigma}_{yy}^\dagger + i(\boldsymbol{\Sigma}_{yx}^\dagger + \boldsymbol{\Sigma}_{xy}^\dagger)\right)$.

The blocks of the $2K \times 2K$ real covariance matrix $\boldsymbol{\Sigma}^\dagger$ are thus given by $\boldsymbol{\Sigma}_{xx}^\dagger = \frac{1}{2}Re(\boldsymbol{\Sigma}_z^\dagger + \tilde{\boldsymbol{\Sigma}}_z^\dagger)$, $\boldsymbol{\Sigma}_{xy}^\dagger = \frac{1}{2}Im(-\boldsymbol{\Sigma}_z^\dagger + \tilde{\boldsymbol{\Sigma}}_z^\dagger)$, $\boldsymbol{\Sigma}_{yx}^\dagger = \frac{1}{2}Im(\boldsymbol{\Sigma}_z^\dagger + \tilde{\boldsymbol{\Sigma}}_z^\dagger)$, and $\boldsymbol{\Sigma}_{yy}^\dagger = \frac{1}{2}Re(\boldsymbol{\Sigma}_z^\dagger - \tilde{\boldsymbol{\Sigma}}_z^\dagger)$, and, in general, both the covariance and the pseudo-covariance matrices are required for a complete second-order characterization of the complex random vector. However, the pseudo-covariance matrix $\tilde{\boldsymbol{\Sigma}}_z^\dagger$ is very rarely introduced in literature, and it is explicitly or implicitly assumed to be zero. This characterizes second order circularity, which means that second-order statistics of $\mathbf{z}^\dagger$ and $e^{i\theta}\mathbf{z}^\dagger$ are the same for any angle θ. A complex random vector, for which the property of second order circularity holds, is called (strictly) *proper* (Neeser and Massey 1993). Specifically, $\mathbf{z}^\dagger$ is proper if $\tilde{\boldsymbol{\Sigma}}_z^\dagger = \boldsymbol{\Sigma}_{xx}^\dagger - \boldsymbol{\Sigma}_{yy}^\dagger + i(\boldsymbol{\Sigma}_{xy}^{\dagger'} + \boldsymbol{\Sigma}_{xy}^\dagger) = 0$ and this is verified if $\boldsymbol{\Sigma}_{xx}^\dagger = \boldsymbol{\Sigma}_{yy}^\dagger$ and $\boldsymbol{\Sigma}_{xy}^\dagger = -\boldsymbol{\Sigma}_{xy}^{\dagger'}$, i.e. $\boldsymbol{\Sigma}_{xy}^\dagger$ must be a *skew-symmetric* matrix. Therefore, the covariance matrix of a proper complex random vector, which alone describes the second-order behavior of ς, can be expressed as $\boldsymbol{\Sigma}_z^\dagger = \boldsymbol{\Sigma}_{xx}^\dagger + \boldsymbol{\Sigma}_{yy}^\dagger + i(\boldsymbol{\Sigma}_{yx}^\dagger - \boldsymbol{\Sigma}_{xy}^\dagger) = 2\left(\mathbf{C}_1^\dagger + i\mathbf{C}_2^\dagger\right)$, where $\mathbf{C}_1^\dagger = \boldsymbol{\Sigma}_{xx}^\dagger = \boldsymbol{\Sigma}_{yy}^\dagger$ and $\mathbf{C}_2^\dagger = \boldsymbol{\Sigma}_{yx}^\dagger = -\boldsymbol{\Sigma}_{xy}^\dagger$. Accordingly, the covariance matrix of the real vector $vec(\mathbf{X}^\dagger) = (Re(\mathbf{z}^\dagger)' \; Im(\mathbf{z}^\dagger)')' = (\mathbf{x}^{\dagger'} \; \mathbf{y}^{\dagger'})'$ has the following structure

$$\boldsymbol{\Sigma}^\dagger = \begin{pmatrix} \mathbf{C}_1^\dagger & -\mathbf{C}_2^\dagger \\ \mathbf{C}_2^\dagger & \mathbf{C}_1^\dagger \end{pmatrix}.$$

The probability density function of a proper complex Gaussian vector, $\mathbf{z}^\dagger \sim \mathscr{CN}_K(\boldsymbol{\mu}_z^\dagger, \boldsymbol{\Sigma}_z^\dagger)$, is defined as

$$f_{\mathscr{CN}}(\mathbf{z}^\dagger; \boldsymbol{\mu}_x^\dagger, \boldsymbol{\Sigma}_z^\dagger) = \frac{1}{\pi^K |\boldsymbol{\Sigma}_z^\dagger|} exp\left(-(\mathbf{z}^\dagger - \boldsymbol{\mu}_z^\dagger)^* \boldsymbol{\Sigma}_z^{\dagger^{-1}} (\mathbf{z}^\dagger - \boldsymbol{\mu}_z^\dagger)\right). \tag{2.1}$$

In fact, given a complex covariance structure, the relation between the quadratic forms in real and complex notation is

$$(vec(\mathbf{X}^\dagger) - vec(\boldsymbol{\mu})^\dagger)' \boldsymbol{\Sigma}^{\dagger^{-1}} (vec(\mathbf{X}^\dagger) - vec(\boldsymbol{\mu})^\dagger) = 2(\mathbf{z}^\dagger - \boldsymbol{\mu}_z^\dagger)^* \boldsymbol{\Sigma}_z^{\dagger^{-1}} (\mathbf{z}^\dagger - \boldsymbol{\mu}_z^\dagger).$$

Furthermore, since it can be shown that each eigenvalue, λ_k, $k = 1, \ldots, K$, of $\boldsymbol{\Sigma}_z^\dagger$, corresponds to a pair of eigenvalues of $\boldsymbol{\Sigma}^\dagger$, it follows that $|\boldsymbol{\Sigma}^\dagger| = \prod_{k=1}^K \left(\frac{\lambda_k}{2}\right)^2$, $|\boldsymbol{\Sigma}_z^\dagger| = \prod_{k=1}^K \lambda_k$ and $|\boldsymbol{\Sigma}^\dagger| = 2^{-2K}|\boldsymbol{\Sigma}_z^\dagger|^2$. Hence we have

$$(2\pi)^K |\boldsymbol{\Sigma}^\dagger|^{1/2} = (2\pi)^K (2^{-2K}|\boldsymbol{\Sigma}_z^\dagger|^2)^{1/2} = \pi^K |\boldsymbol{\Sigma}_z^\dagger|.$$

Proper complex zero-mean Gaussian vectors are also called *circularly complex Gaussian* vectors because their pdf is rotationally invariant, that is it remains the same if we rotate each component by some angle θ, so that $e^{i\theta}\mathbf{z}^\dagger$ has the same pdf as $\mathbf{z}^\dagger$. For a non centered random vector, the pdf is unchanged if we rotate each component about its mean by some angle θ.

Particular cases of the complex covariance structure are the cyclic Markov model (Dryden and Mardia 1998, p.139) and the isotropic covariance case. The cyclic Markov covariance structure is a particular case of the complex covariance model and can be useful for modeling the covariance structure of landmark "regularly" scattered on a closed outline. It is characterized by a block diagonal structure, $\boldsymbol{\Sigma}^\dagger = \sigma^2\mathbf{I}_2 \otimes \mathbf{R}_1$, where the entries of $\mathbf{R}_1$ are defined as

$$\mathbf{R}_1(k,l) = \frac{\vartheta^{|k-l|} + \vartheta^{K-|k-l|}}{1-\vartheta^K} \qquad 1 \leq k,l \leq K \quad and \quad 0 \leq \vartheta < 1.$$

The isotropic covariance structure is also a particular case of the cyclic Markov model for which $\mathbf{R}_1 = \mathbf{I}_K$; therefore the covariance matrix is given by $\boldsymbol{\Sigma}^\dagger = \sigma^2\mathbf{I}_{2K}$.

2.3 The Gaussian Distribution in the Pre-form Space

For shape registration purposes, the information about location need to be removed and, considering Bookstein coordinates introduced in Sect. 1.3.2, the preform of the configuration $\mathbf{X}^\dagger$ is obtained by the matrix transformation $\mathbf{X} = \mathbf{L}\mathbf{X}^\dagger$. Here $\mathbf{L}$ is a $(K-1) \times K$ translation matrix constructed as $(-\mathbf{1}_{K-1}, \mathbf{I}_{K-1})$, where $\mathbf{I}_{K-1}$ is the identity matrix of dimension $(K-1)A \times (K-1)$ and $\mathbf{1}_{K-1}$ is a $(K-1)$-dimensional vector of ones.

Since this transformation represents a linear projection from $\Re^{2K}$ to $\Re^{2K-1}$, the pdf in the pre-form space is Gaussian with mean $\boldsymbol{\mu} = \mathbf{L}\boldsymbol{\mu}^\dagger$ and covariance matrix $\boldsymbol{\Sigma} = (\mathbf{I}_2 \otimes \mathbf{L}_2)\boldsymbol{\Sigma}^\dagger(\mathbf{I} \otimes \mathbf{L}')$. It thus follows that $vec(\mathbf{X}) \sim \mathcal{N}_{2K-2}(vec(\boldsymbol{\mu}), \boldsymbol{\Sigma})$.

In complex notation, the landmark coordinates in the complex preform space are given by $\mathbf{z} = \mathbf{L}\mathbf{z}^\dagger$, with $\mathbf{z} = \mathbf{x} + i\mathbf{y}$. Accordingly, $\mathbf{z} \sim \mathcal{CN}_{K-1}(\boldsymbol{\mu}_z, \boldsymbol{\Sigma}_z)$, where $\boldsymbol{\mu}_z = \mathbf{L}\boldsymbol{\mu}_z^\dagger$ and $\boldsymbol{\Sigma}_z = \mathbf{L}\boldsymbol{\Sigma}_z^\dagger\mathbf{L}'$.

2.4 The Offset-Normal Shape Distribution

The shape space of centered configurations is obtained by removing the information about rotation and scaling. Here, without loss of generality, we work with Bookstein shape coordinates, $\mathbf{U}$, which are based on the transformation

$$\mathbf{X} \rightarrow \mathbf{U} = \varphi \mathbf{X}\mathbf{R} = \frac{1}{x_2^2 + y_2^2} \begin{pmatrix} x_2 & y_2 \\ x_3 & y_3 \\ \vdots & \vdots \\ x_K & y_K \end{pmatrix} \begin{pmatrix} x_2 & -y_2 \\ y_2 & x_2 \end{pmatrix} = \begin{pmatrix} 1 & 0 \\ u_3 & v_3 \\ \vdots & \vdots \\ u_K & v_K \end{pmatrix}.$$

Given the pdf of landmark coordinates in the pre-form space

$$f\,(vec(\mathbf{X}); \boldsymbol{\mu}, \boldsymbol{\Sigma}) = \frac{1}{(2\pi)^{K-1}|\boldsymbol{\Sigma}|^{\frac{1}{2}}} exp \left\{ -\frac{1}{2}[vec(\mathbf{X}) - vec(\boldsymbol{\mu})]' \right.$$
$$\left. \boldsymbol{\Sigma}^{-1}[vec(\mathbf{X}) - vec(\boldsymbol{\mu})] \right\}, \tag{2.2}$$

the distribution of the shape random variables, $\mathbf{u} = \{u_k, v_k\}_{k=3:K}$, is obtained by integrating out the information on scale and rotation parameters.
Given the matrix

$$\mathbf{W} = \begin{pmatrix} 1 & u_3 & \dots & u_K & 0 & v_3 & \dots & v_K \\ 0 & -v_3 & \dots & -v_K & 1 & u_3 & \dots & u_K \end{pmatrix}'$$

and the vector $\mathbf{h} = (x_2\ y_2)'$, representing the rotation and scale information in the preform space, the vectorial form of the preform configuration can equivalently be expressed as $vec(\mathbf{X}) = \mathbf{W}\mathbf{h}$. Therefore, the joint distribution of $(\mathbf{h}, \mathbf{u})$ can be written as

$$f\,(\mathbf{h}, \mathbf{u}; \boldsymbol{\mu}, \boldsymbol{\Sigma}) = \frac{1}{(2\pi)^{K-1}|\boldsymbol{\Sigma}|^{\frac{1}{2}}} exp \left\{ -\frac{\left[\mathbf{W}\mathbf{h} - vec(\boldsymbol{\mu})]'\boldsymbol{\Sigma}^{-1}[\mathbf{W}\mathbf{h} - vec(\boldsymbol{\mu})\right]}{2} \right\}$$
$$|J(\mathbf{X} \rightarrow (\mathbf{h}, \mathbf{u}))|$$

where $|J(\mathbf{X} \rightarrow (\mathbf{h}, \mathbf{u}))| = ||\mathbf{h}||^{2(K-2)}$ is the Jacobian of the transformation $\mathbf{X} \rightarrow (\mathbf{h}, \mathbf{u})$.

Setting $\boldsymbol{\Gamma} = \left(\mathbf{W}'\boldsymbol{\Sigma}^{-1}\mathbf{W}\right)^{-1}$, $\boldsymbol{\upsilon} = \boldsymbol{\Gamma}\mathbf{W}'\boldsymbol{\Sigma}^{-1}vec(\boldsymbol{\mu})$, and $g = vec(\boldsymbol{\mu})'\boldsymbol{\Sigma}^{-1}vec(\boldsymbol{\mu}) - \boldsymbol{\upsilon}'\boldsymbol{\Gamma}^{-1}\boldsymbol{\upsilon}$, the joint distribution can be re-expressed in terms of the random vector of rotation and scale parameters $\mathbf{h}$

$$f\,(\mathbf{h}, \mathbf{u}; \boldsymbol{\mu}, \boldsymbol{\Sigma}) = \frac{exp(-g/2)}{(2\pi)^{K-1}|\boldsymbol{\Sigma}|^{\frac{1}{2}}} exp \left\{ -\frac{(\mathbf{h} - \boldsymbol{\upsilon})'\boldsymbol{\Gamma}^{-1}(\mathbf{h} - \boldsymbol{\upsilon})}{2} \right\} ||\mathbf{h}||^{2(K-2)}. \tag{2.3}$$

Equation (2.3) can be simplified by considering the eigen-decomposition $\boldsymbol{\Gamma} = \boldsymbol{\Psi}\mathbf{D}\boldsymbol{\Psi}'$, where $\boldsymbol{\Psi}$ is a 2×2 eigenvector matrix and $\mathbf{D} = diag(\sigma_1^2, \sigma_2^2)$ is the diagonal matrix of the corresponding eigenvalues. In the new coordinate system, the scale and rotation parameters and their mean vector are given by $\mathbf{l} = \boldsymbol{\Psi}'\mathbf{h}$ and $\boldsymbol{\zeta} = \boldsymbol{\Psi}'\boldsymbol{\upsilon}$, respectively. Accordingly, the quadratic form in Eq. (2.3) can be rewritten as $(\mathbf{h}-\boldsymbol{\upsilon})'\boldsymbol{\Gamma}^{-1}(\mathbf{h}-\boldsymbol{\upsilon}) = (\mathbf{l}-\boldsymbol{\zeta})'\mathbf{D}^{-1}(\mathbf{l}-\boldsymbol{\zeta}) = \frac{(l_1-\zeta_1)^2}{\sigma_1^2} + \frac{(l_2-\zeta_2)^2}{\sigma_2^2}$, and the jacobian of the transformation expressed as $\|\mathbf{h}\|^{2(K-2)} = (\mathbf{h}'\mathbf{h})^{K-2} = (\mathbf{h}'\boldsymbol{\Psi}\boldsymbol{\Psi}'\mathbf{h})^{K-2} = (\mathbf{l}'\mathbf{l})^{K-2} = (l_1^2+l_2^2)^{K-2}$. The joint pdf of $(\mathbf{l},\mathbf{u})$ can now be formulated as

$$\begin{aligned} f(\mathbf{l},\mathbf{u};\boldsymbol{\mu},\boldsymbol{\Sigma}) &= \frac{exp(-g/2)}{(2\pi)^{K-1}|\boldsymbol{\Sigma}|^{\frac{1}{2}}} exp\left(\frac{l_1-\zeta_1}{\sigma_1}\right)^2 exp\left(\frac{l_2-\zeta_2}{\sigma_2}\right)^2 (l_1^2+l_2^2)^{K-2} \\ &= \frac{|\boldsymbol{\Gamma}|^{\frac{1}{2}} exp(-g/2)}{(2\pi)^{K-2}|\boldsymbol{\Sigma}|^{\frac{1}{2}}} f_{\mathcal{N}}(l_1;\zeta_1,\sigma_1) f_{\mathcal{N}}(l_2;\zeta_2,\sigma_2) \sum_{j=0}^{K-2}\binom{K-2}{j} l_1^{2j} l_2^{2(K-2-j)} \end{aligned} \tag{2.4}$$

where we have used the binomial expansion $(l_1^2+l_2^2)^{K-2} = \sum_{j=0}^{K-2}\binom{K-2}{j} l_1^{2j} l_2^{2(K-2-j)}$.

The marginal distribution of the shape variables can be obtained by integrating out the scale and rotation parameters

$$\begin{aligned} f(\mathbf{u};\boldsymbol{\mu},\boldsymbol{\Sigma}) &= \frac{|\boldsymbol{\Gamma}|^{\frac{1}{2}} exp(-g/2)}{(2\pi)^{K-2}|\boldsymbol{\Sigma}|^{\frac{1}{2}}} \sum_{j=0}^{K-2}\binom{K-2}{j} \int l_1^{2j} f_{\mathcal{N}}(l_1;\zeta_1,\sigma_1)dl_1 \\ &\quad \int l_2^{2(K-2-j)} f_{\mathcal{N}}(l_2;\zeta_2,\sigma_2)dl_2 \\ &= \frac{|\boldsymbol{\Gamma}|^{\frac{1}{2}} exp(-g/2)}{(2\pi)^{K-2}|\boldsymbol{\Sigma}|^{\frac{1}{2}}} \sum_{j=0}^{K-2}\binom{K-2}{j} E[l_1^{2j}|\zeta_1,\sigma_1] E[l_2^{2(K-2-j)}|\zeta_2,\sigma_2] \end{aligned} \tag{2.5}$$

where $E[l^p|\zeta,\sigma]$ denotes the moments of the univariate Gaussian distribution with parameters (ζ,σ). These moments are calculated as (see 3.462/4 and 8.972 in Gradshteyn and Ryzhik 1980)

$$E[l^p|\zeta,\sigma] = \begin{cases} (2\sigma^2)^q q! L_q^{(-1/2)}\left(\frac{-\zeta^2}{2\sigma^2}\right) & if\ p = 2q \\ \zeta(2\sigma^2)^q q! L_q^{(1/2)}\left(\frac{-\zeta^2}{2\sigma^2}\right) & if\ p = 2q+1 \end{cases} \tag{2.6}$$

where

$$L_q^{(\alpha)}(x) = \sum_{j=1}^{q} \frac{(1+\alpha)_q(-x)^j}{(1+\alpha)_j j!(q-j)!}$$

is the generalized Laguerre polynomial of order q.

2.5 EM Algorithm for Estimating $\boldsymbol{\mu}$ and $\boldsymbol{\Sigma}$

The Expectation-Maximization (EM) algorithm (Dempster et al. 1977) is a general iterative procedure that attempts to find the maximum likelihood estimators of some unknown parameters in the presence of "missing" or "hidden" data.

The EM algorithm for the estimation of the parameters of the offset-normal shape distribution was first proposed by Kume and Welling (2010).

Given a random sample of N configurations, let $\mathscr{X} = \{\mathbf{X}^{(n)}\}_{n=1:N}$ and $\mathscr{U} = \{\mathbf{u}^{(n)}\}_{n=1:N}$ denote the full data in the preform space and the observed data in the shape space, respectively. Also, let $l_0(\boldsymbol{\mu}, \boldsymbol{\Sigma}|\mathscr{X}) = \sum_{n=1}^{N} logf_{\mathscr{N}}(\mathbf{X}^{(n)}|\boldsymbol{\mu}, \boldsymbol{\Sigma})$ be the complete data log-likelihood and $l(\boldsymbol{\mu}, \boldsymbol{\Sigma}|\mathscr{U}) = \sum_{n=1}^{N} logf(\mathbf{u}^{(n)}|\boldsymbol{\mu}, \boldsymbol{\Sigma})$ be the shape data log-likelihood. Here, as discussed in the previous sections, $vec(\mathbf{X}^{(n)}) \sim \mathscr{N}_{2(K-1)}(\boldsymbol{\mu}, \boldsymbol{\Sigma})$, and $f(\mathbf{u}^{(n)}|\boldsymbol{\mu}, \boldsymbol{\Sigma})$ is the induced pdf of shape variables $\boldsymbol{u}^{(n)}$.

As highlighted by Kume and Welling (2010), due to shape invariance with respect to scaling and rotation of pre-forms, it is possible to estimate in terms of $\mathscr{U}$ only those parameters which identify the equivalent class

$$\boldsymbol{\Theta} = \left\{\left[\varphi\boldsymbol{\mu}\mathbf{R}, s^2(\mathbf{R}' \otimes \mathbf{I}_{K-1})\boldsymbol{\Sigma}(\mathbf{R} \otimes \mathbf{I}_{K-1})\right] |\varphi \in \Re^+, \mathbf{R} \in SO(2)\right\}. \tag{2.7}$$

Therefore, the proposed estimation method deals with only those parameters which identify this equivalent class and not all those identifying the mean and the covariance matrix in the configuration space. Hence, the target is to find the values of $\boldsymbol{\mu}$ and $\boldsymbol{\Sigma}$, identifying equivalent classes, which maximize the log-likelihood function $l(\boldsymbol{\mu}, \boldsymbol{\Sigma}|\mathscr{U})$.

Regarding the number of parameters to be estimated, we notice that, while in the configuration space there are $2K$ parameters for the mean vector, and $K(2K + 1)$ parameters for the covariance matrix, in the preform space we have $2(K - 1)$ parameters for the mean vector, and $(K - 1)(2(K - 1) + 1)$ parameters for the covariance matrix. Therefore, at most $2(K-1) + (K-1)(2K-1)$ parameters could be identified. Without loss of generality we can assume that the mean $\boldsymbol{\mu}$ is re-scaled and rotated such that its first landmark is $(1, 0)$. So there are at most $2(K - 2)$ parameters for the mean and $(K - 1)(2K - 1) = 2K^2 - 3K + 1$ for the covariance matrix identifying $\boldsymbol{\Theta}$ in the equivalence classes (2.7). While the parameters for the shape of $\boldsymbol{\mu}$ are fully identifiable, Dryden and Mardia (1998, p. 138) expect that only $(K - 2)(2K - 3) = 2K^2 - 7K + 6$ parameters are practically identifiable for $\boldsymbol{\Sigma}$. However, certain conditions on $\boldsymbol{\Sigma}$ avoid this identification problem. In particular, the entries of a complex covariance structure, $\boldsymbol{\Sigma}_z = 2(\mathbf{C}_1 + i\mathbf{C}_2)$, are fully identifiable up to some re-scaling constant. In fact, since $\mathbf{C}_1$ is symmetric and has $(K-1)K/2$ parameters, while $\mathbf{C}_2$ is skew-symmetric and has $(K - 1)K/2 - (K-1)$ parameters, $\boldsymbol{\Sigma}_z$ has $(K - 1)^2 = K^2 - 2K + 1$ parameters to be estimated.

Since it is simpler to maximize the complete data log-likelihood $l_0(\boldsymbol{\mu}, \boldsymbol{\Sigma}|\mathscr{X})$, rather than working with the shape data log-likelihood, $l(\boldsymbol{\mu}, \boldsymbol{\Sigma}|\mathscr{U})$, Kume and Welling (2010) propose to exploit the EM algorithm which attempts to maximize the complete data log-likelihood iteratively, by replacing $l_0(\boldsymbol{\mu}, \boldsymbol{\Sigma}|\mathscr{X})$ by its conditional expectation given the observed data $\mathscr{U}$. This expectation, denoted as $\mathscr{Q}_{\boldsymbol{\mu}^{(r)},\boldsymbol{\Sigma}^{(r)}}(\boldsymbol{\mu},\boldsymbol{\Sigma})$, is computed with respect to the distribution of the full data, given the observed $\mathscr{U}$, and is evaluated at the current parameter estimates, $\boldsymbol{\mu}^{(r)}$ and $\boldsymbol{\Sigma}^{(r)}$.

Thus the EM algorithm consists of an estimation step (E-step) followed by a maximization step (M-step) and these are defined as

- **E-step**: compute $\mathscr{Q}_{\boldsymbol{\mu}^{(r)},\boldsymbol{\Sigma}^{(r)}}(\boldsymbol{\mu},\boldsymbol{\Sigma}) = \mathbf{E}_{\mathscr{X}|\mathscr{U},\boldsymbol{\mu}^{(r)},\boldsymbol{\Sigma}^{(r)}}[l_0(\boldsymbol{\mu},\boldsymbol{\Sigma}|\mathscr{X})]$;
- **M-step**: find $\boldsymbol{\mu}^{(r+1)}$, $\boldsymbol{\Sigma}^{(r+1)}$ such that $\mathscr{Q}_{\boldsymbol{\mu}^{(r+1)},\boldsymbol{\Sigma}^{(r+1)}}(\boldsymbol{\mu},\boldsymbol{\Sigma}) \geq \mathscr{Q}_{\boldsymbol{\mu}^{(r)},\boldsymbol{\Sigma}^{(r)}}(\boldsymbol{\mu},\boldsymbol{\Sigma})$.

Under the assumption of normality and given N independently, identically distributed random samples, the conditional expected log-likelihood is given by

$$\mathscr{Q}_{\boldsymbol{\mu}^{(r)},\boldsymbol{\Sigma}^{(r)}}(\boldsymbol{\mu},\boldsymbol{\Sigma}) = \sum_{n=1}^{N}\int log\big(f_{\mathscr{N}}(\mathbf{X}^{(n)}|\boldsymbol{\mu},\boldsymbol{\Sigma})\big)dF(\mathbf{X}^{(n)}|\mathbf{u}^{(n)},\boldsymbol{\mu}^{(r)},\boldsymbol{\Sigma}^{(r)}), \tag{2.8}$$

where $dF(\mathbf{X}^{(n)}|\mathbf{u}^{(n)},\boldsymbol{\mu}^{(r)},\boldsymbol{\Sigma}^{(r)})$ is the conditional distribution of $\mathbf{X}^{(n)}$ given its shape $\mathbf{u}^{(n)}$. The updated values can be calculated once we know how to maximize the conditional expected log-likelihood. The computation of the E-step and the M-step simplifies when the distribution of the full-data belongs to the exponential family. In this case, the E-step reduces to computing the expectation of the complete data sufficient statistics given the observed data at current parameter estimates. The M-step involves maximizing the expected log-likelihood computed in the E-step. In the exponential family case, maximizing the expected log-likelihood to obtain the next iterate can be avoided. Instead, the conditional expectations of the sufficient statistics computed in the E-step can be directly substituted for the sufficient statistics that occur in the expressions obtained for the complete-data maximum likelihood estimators, to obtain the next iterate. Hence, the two steps appear as follows

M-step: the maximum of $Q_{\boldsymbol{\mu}^{(r)}),\boldsymbol{\Sigma}^{(r)})}(\boldsymbol{\mu},\boldsymbol{\Sigma})$ is achieved at

$$vec(\boldsymbol{\mu}^{(r+1)}) = \frac{1}{N}\sum_{n=1}^{N}\int vec(\mathbf{X}^{(n)})dF(\mathbf{X}^{(n)}|\mathbf{u}^{(n)},\boldsymbol{\mu}^{(r)},\boldsymbol{\Sigma}^{(r)})$$

$$\begin{aligned}\boldsymbol{\Sigma}^{(r+1)} &= \frac{1}{N}\sum_{n=1}^{N}\int vec(\mathbf{X}^{(n)})vec(\mathbf{X}^{(n)})'dF(\mathbf{X}^{(n)}|\mathbf{u}^{(n)},\boldsymbol{\mu}^{(r)},\boldsymbol{\Sigma}_r)\\ &\quad -vec(\boldsymbol{\mu}^{(r+1)})vec(\boldsymbol{\mu}^{(r+1)}))'\end{aligned}$$

E-step: the complete data sufficient statistics, given the observed data and current parameter estimates, can be computed as in Lemma 1 in Kume and Welling (2010), that is

$$\int vec(\mathbf{X})dF(\mathbf{X}|\mathbf{u},\boldsymbol{\mu},\boldsymbol{\Sigma}) = \mathbf{W}\boldsymbol{\Psi}\frac{\int \mathbf{l}f(\mathbf{l},\mathbf{u};\boldsymbol{\mu},\boldsymbol{\Sigma})d\mathbf{l}}{f(\mathbf{u};\boldsymbol{\mu},\boldsymbol{\Sigma})} \tag{2.9}$$

and

$$\int vec(\mathbf{X})vec(\mathbf{X})'dF(\mathbf{X}|\mathbf{u},\boldsymbol{\mu},\boldsymbol{\Sigma}) = \mathbf{W}\boldsymbol{\Psi}\frac{\int \mathbf{l}\mathbf{l}'f(\mathbf{l},\mathbf{u};\boldsymbol{\mu},\boldsymbol{\Sigma})d\mathbf{l}}{f(\mathbf{u};\boldsymbol{\mu},\boldsymbol{\Sigma})}\boldsymbol{\Psi}'\mathbf{W}'. \tag{2.10}$$

Notice that for the pairs $(a, b) \in \{(1,0),(0,1),(2,0),(1,1),(0,2)\}$, we have

$$\frac{\int l_1^a l_2^b f(\mathbf{l},\mathbf{u};\boldsymbol{\mu},\boldsymbol{\Sigma})d\mathbf{l}}{f(\mathbf{u};\boldsymbol{\mu},\boldsymbol{\Sigma})} = \frac{\sum_{j=0}^{K-2}\binom{K-2}{j}E[l_1^{2j+a}|\zeta_1,\sigma_1]E[l_2^{2(K-2-j)+b}|\zeta_2,\sigma_2]}{\sum_{j=0}^{K-2}\binom{K-2}{j}E[l_1^{2j}|\zeta_1,\sigma_1]E[l_2^{2(K-2-j)}|\zeta_2,\sigma_2]} \tag{2.11}$$

where the expectations can be computed as shown in Eq. (2.6). Proofs of Eq. (2.11) are provided in Appendix A.1.

2.5.1 EM for Complex Covariance

As discussed in the previous section, identifiability problems can be solved by considering a complex structure for the covariance matrix. In fact, the proper complex normal distribution is particularly important since the covariance matrix parameters are fully identifiable. Furthermore, for a proper complex random vector, the covariance structure remains invariant under rotations.

In the preform space, as shown in Sect. 2.3, the pdf of the $(K-1)$-dimensional complex vector $\mathbf{z} = \mathbf{x} + i\mathbf{y}$, with mean $\boldsymbol{\mu}_z = \boldsymbol{\mu}_x + i\boldsymbol{\mu}_y$ and complex covariance $\boldsymbol{\Sigma}_z = 2(\mathbf{C}_1 + i\mathbf{C}_2)$, is given by

$$f_{\mathscr{C}\mathscr{N}}\left(\mathbf{z};\boldsymbol{\mu}_z,\boldsymbol{\Sigma}_z\right) = \frac{1}{\pi^{(K-1)}\left|\boldsymbol{\Sigma}_z\right|}\exp\left\{-\left(\mathbf{z}-\boldsymbol{\mu}_z\right)^*\boldsymbol{\Sigma}_z^{-1}\left(\mathbf{z}-\boldsymbol{\mu}_z\right)\right\}. \tag{2.12}$$

For a complex coordinate system, removing the rotation and scaling parameters is easy. In fact, the rotated configuration is obtained as $\boldsymbol{\xi} = \mathbf{z}/z_2$, where $\xi_k = u_k + iv_k$, $k = 2,\ldots,K$, and $z_2 = x_2 + iy_2$ represent the scale and rotation parameters. Given the Jacobian $\|z_2\|^{2(K-2)}$ of the transformation $\mathbf{z} \to (z_2, \boldsymbol{\xi})$, the joint pdf can be expressed as

$$f_{\mathscr{C}\mathscr{N}}\left(\boldsymbol{\xi},z_2;\boldsymbol{\mu}_z,\boldsymbol{\Sigma}_z\right) = \frac{1}{\pi^{(K-1)}\left|\boldsymbol{\Sigma}_z\right|}\exp\left\{-\left(\boldsymbol{\xi}z_2-\boldsymbol{\mu}_z\right)^*\boldsymbol{\Sigma}_z^{-1}\left(\boldsymbol{\xi}z_2-\boldsymbol{\mu}_z\right)\right\}\|z_2\|^{2(K-2)}. \tag{2.13}$$

Setting $\gamma_z = \left(\boldsymbol{\xi}^*\boldsymbol{\Sigma}_z^{-1}\boldsymbol{\xi}\right)^{-1}$, $\eta = \gamma_z\boldsymbol{\xi}^*\boldsymbol{\Sigma}_z^{-1}\boldsymbol{\mu}_z$, and $g_z = \boldsymbol{\mu}_z^*\boldsymbol{\Sigma}_z^{-1}\boldsymbol{\mu}_z - \bar{\eta}\gamma_z^{-1}\eta$, Eq. (2.13) can be rewritten as

$$f_{\mathscr{C}\mathscr{N}}\left(\boldsymbol{\xi},z_2;\boldsymbol{\mu}_z,\boldsymbol{\Sigma}_z\right) = \frac{\exp\{-g_z\}}{\pi^{(K-1)}\left|\boldsymbol{\Sigma}_z\right|}\exp\left\{-\left(\bar{z}_2-\bar{\eta}\right)\gamma_z^{-1}\left(z_2-\eta\right)\right\}\|z_2\|^{2(K-2)}$$

which now expresses the pdf in terms of the scale and rotation parameters. Since any subvector of a proper random vector is also proper (Neeser and Massey 1993), the random variable z_2 has a complex Gaussian distribution with mean η and variance γ_z. Therefore the joint pdf of Eq. (2.13) can be also written as

$$f_{\mathscr{CN}}\left(\boldsymbol{\xi}, z_2; \boldsymbol{\mu}_z, \boldsymbol{\Sigma}_z\right) = \frac{\gamma_z \exp\{-g_z\}}{\pi^{(K-2)}\left|\boldsymbol{\Sigma}_z\right|} \|z_2\|^{2(K-2)} f_{\mathscr{CN}}\left(z_2; \eta, \gamma_z\right). \tag{2.14}$$

Kume and Welling (2010) show that the updated values $\boldsymbol{\mu}_z^{(r+1)}$ and $\boldsymbol{\Sigma}_z^{(r+1)}$, obtained in the M-step of the EM algorithm, are

$$\begin{aligned}\boldsymbol{\mu}_z^{(r+1)} &= \frac{1}{N}\sum_{n=1}^{N}\int \mathbf{z}^{(n)} dF\left(\mathbf{z}^{(n)}|\boldsymbol{\xi}^{(n)}, \boldsymbol{\mu}_z^{(r)}, \boldsymbol{\Sigma}_z^{(r)}\right) = \\ &= \frac{1}{N}\sum_{n=1}^{N}\boldsymbol{\xi}^{(n)}\frac{\int z_2 \|z_2\|^{2(K-2)} f_{\mathscr{CN}}\left(z_2, \boldsymbol{\xi}^{(n)}; \boldsymbol{\mu}_z^{(r)}, \boldsymbol{\Sigma}_z^{(r)}\right) dz_2}{\int \|z_2\|^{2(K-2)} f_{\mathscr{CN}}\left(z_2, \boldsymbol{\xi}^{(n)}; \boldsymbol{\mu}_z^{(r)}, \boldsymbol{\Sigma}_z^{(r)}\right) dz_2}\end{aligned} \tag{2.15}$$

and

$$\begin{aligned}\boldsymbol{\Sigma}_z^{(r+1)} &= \frac{1}{N}\sum_{n=1}^{N}\int \mathbf{z}^{(n)}\mathbf{z}^{(n)*} dF\left(\mathbf{z}^{(n)}|\boldsymbol{\xi}^{(n)}, \boldsymbol{\mu}_z^{(r)}, \boldsymbol{\Sigma}_z^{(r)}\right) - \boldsymbol{\mu}_z^{(r+1)}\boldsymbol{\mu}_z^{(r+1)*} = \\ &= \frac{1}{N}\sum_{n=1}^{N}\boldsymbol{\xi}^{(n)}\boldsymbol{\xi}^{(n)*}\frac{\int \|z_2\|^{2(K-1)} f_{\mathscr{CN}}\left(z_2, \boldsymbol{\xi}^{(n)}; \boldsymbol{\mu}_z^{(r)}, \boldsymbol{\Sigma}_z^{(r)}\right) dz_2}{\int \|z_2\|^{2(K-2)} f_{\mathscr{CN}}\left(z_2, \boldsymbol{\xi}^{(n)}; \boldsymbol{\mu}_z^{(r)}, \boldsymbol{\Sigma}_z^{(r)}\right) dz_2} \\ &\quad -\boldsymbol{\mu}_z^{(r+1)}\boldsymbol{\mu}_z^{(r+1)*}\end{aligned} \tag{2.16}$$

with ratios calculated as

$$\frac{\int z_2 \|z_2\|^{2(K-2)} f_{\mathscr{CN}}\left(z_2, \boldsymbol{\xi}; \boldsymbol{\mu}_z, \boldsymbol{\Sigma}_z\right) dz_2}{\int \|z_2\|^{2(K-2)} f_{\mathscr{CN}}\left(z_2, \boldsymbol{\xi}; \boldsymbol{\mu}_z, \mathbf{C}\right) dz_2} = \frac{\omega\,(K-1)}{\|\eta\|}\left(\frac{\mathscr{L}_{K-1}\left(-\|\eta\|^2/\gamma_z\right)}{\mathscr{L}_{K-2}\left(-\|\eta\|^2/\gamma_z\right)} - 1\right) \tag{2.17}$$

and

$$\frac{\int \|z_2\|^{2(K-1)} f_{\mathscr{CN}}\left(z_2, \boldsymbol{\xi}; \boldsymbol{\mu}_z, \boldsymbol{\Sigma}_z\right) dz_2}{\int \|z_2\|^{2(K-2)} f_{\mathscr{CN}}\left(z_2, \boldsymbol{\xi}; \boldsymbol{\mu}_z, \boldsymbol{\Sigma}_z\right) dz_2} = \gamma_z\,(K-1)\left(\frac{\mathscr{L}_{K-1}\left(-\|\eta\|^2/\gamma_z\right)}{\mathscr{L}_{K-2}\left(-\|\eta\|^2/\gamma_z\right)}\right) \tag{2.18}$$

where $\omega = e^{-i\theta}$, such that $\bar{\omega}\boldsymbol{\xi}^*\boldsymbol{\Sigma}_z^{-1}\boldsymbol{\mu}_z$ is a positive number. Proofs for Eqs. (2.17) and (2.18) are provided in Appendix A.2.

2.5.2 Cyclic Markov and Isotropic Covariances

Assume that the complex covariance follows the cyclic Markov model, i.e. $\mathbf{C}_2^\dagger = \mathbf{0}$ and $\mathbf{C}_1^\dagger = \sigma^2 \mathbf{R}_1$. Since the estimation is based on identifying elements from the equivalence class (2.7), we can assume $\sigma^2 = 1/2$ and the estimation of the mean vector and the covariance parameter ϑ can be estimated through a generalized EM algorithm (McLachlan and Krishnan 1997).

More specifically, given an initial estimate of the correlation parameter $\vartheta_{(r)}$, and denoting with $\mathbf{R}_1^\dagger(k,l) = \frac{\vartheta_{(r)}^{|k-l|} + \vartheta_{(r)}^{K-|k-l|}}{1-\vartheta_{(r)}^K}$ (see Sect. 2.2) the entries of $\mathbf{R}_1^\dagger$, it is possible to construct the covariance matrix in the pre-form space as $\boldsymbol{\Sigma}_z^{(r)} = \mathbf{L}\mathbf{R}_1^\dagger \mathbf{L}'$, and compute the updated estimate of the mean vector, $\boldsymbol{\mu}_z^{(r+1)}$, as in Eq. (2.15). Then, replacing $\boldsymbol{\mu}_z^{(r+1)}$ in the conditional log-likelihood,

$$\mathcal{Q}_{\boldsymbol{\mu}_z^{(r)}, \boldsymbol{\Sigma}_z^{(r)}}\left(\boldsymbol{\eta}_{r+1}, \boldsymbol{\Sigma}_z\right) = -N \log |\boldsymbol{\Sigma}_z| - \sum_{n=1}^{N} \int \left(\mathbf{z}^{(n)} - \boldsymbol{\mu}_z^{(r+1)}\right)^* \boldsymbol{\Sigma}_z^{-1} \left(\mathbf{z}^{(n)} - \boldsymbol{\mu}_z^{(r+1)}\right) dF\left(\mathbf{z}^{(n)} | \boldsymbol{\xi}^{(n)}, \boldsymbol{\mu}_z^{(r)}, \boldsymbol{\Sigma}_z^{(r)}\right)$$

and noting that $\left(\mathbf{z} - \boldsymbol{\mu}_z\right)^* \boldsymbol{\Sigma}_z^{-1} \left(\mathbf{z} - \boldsymbol{\mu}_z\right) = tr\left(\boldsymbol{\Sigma}_z^{-1} \left(\mathbf{z} - \boldsymbol{\mu}_z\right)\left(\mathbf{z} - \boldsymbol{\mu}_z\right)^*\right)$, we thus find the value of ϑ_{r+1} which maximizes

$$\mathcal{Q}_{\boldsymbol{\mu}_z^{(r)}, \boldsymbol{\Sigma}_z^{(r)}}\left(\boldsymbol{\eta}_{r+1}, \boldsymbol{\Sigma}_z\right) = -N \log |\boldsymbol{\Sigma}_z| - tr\left\{\boldsymbol{\Sigma}_z^{-1}\left[\sum_{n=1}^{N} \int \mathbf{z}^{(n)*} \mathbf{z}^{(n)} dF\left(\mathbf{z}^{(n)} | \boldsymbol{\xi}^{(n)}, \boldsymbol{\mu}_z^{(r)}, \boldsymbol{\Sigma}_z^{(r)}\right) - N \boldsymbol{\mu}_z^{(r+1)*} \boldsymbol{\mu}_z^{(r+1)}\right]\right\}.$$

Since the values of $\int \mathbf{z}^{(n)*} \mathbf{z}^{(n)} dF\left(\mathbf{z}^{(n)} | \boldsymbol{\xi}^{(n)}, \boldsymbol{\mu}_z^{(r)}, \boldsymbol{\Sigma}_z^{(r)}\right)$ are obtained as in Eq. (2.16), this is a simple univariate optimization problem which can be carried out numerically.

If we consider an isotropic covariance structure, $\boldsymbol{\Sigma}_z^\dagger = 2\sigma^2 \mathbf{I}_k$, since this corresponds to the assumption of uncorrelated landmarks, it is possible to set $\sigma^2 = 1/2$, and it suffices for the EM algorithm now to calculate only $\boldsymbol{\mu}_z^{(r+1)}$.

2.6 Data Analysis: The FG-NET Data

In this section we introduce the FG-NET (Face and Gesture Recognition Research Network) database with facial expressions and emotions from the Technical University Munich (Wallhoff 2006). The data set has been generated in an attempt to assist researchers who investigate the effects of different facial expressions as part of the European Union project. This is an image database containing face images showing a number of subjects performing the six different expressions defined by Ekman and Friesen (1971). Here, we mainly focus on *happiness* and *surprise* expressions for which video sequences are available. The dynamics of the two expressions is described by the changes in time of the landmark coordinates. All acquired sequences in the FG-NET database are starting from the neutral state passing into the emotional state. Depending on the expression, a single recorded sequence can take up to several seconds. For each subject, a transcription of the start, apex, and hold frame (i.e. up to which frame it is possible to see the emotion) can be found with the metadata file made available by the Interactive System Group (ISG, Technical University of Munich). Since, in average, about 20 frames separate the start from the apex, only a few frames can be used to describe the dynamics of the complete expression. In our case, we work with 7 frames chosen at equally spaced intervals. For each individual, the first available time frame represents the neutral expression which, in our analysis, was used as the reference configuration to estimate the pole of the tangent projections. For each frame, we then consider the material gathered from 16 different individuals and summarize the expressions (at frame $t = 1, 2, \ldots, 7$) through a set of 34 landmarks manually placed on the face of each subject.

The facial landmark configuration is shown in Fig. 2.1 where the selected points are used to represent the eyebrows (10 landmarks), the eyelid margins (16 landmarks) and the mouth region (8 landmarks).

By following the likelihood approach, we start with an experiment in which we first estimate the mean shapes of the two expressions. Then, in order to test their differences, we compare both *happiness* and *surprise* expressions under isotropic and complex covariance structures. Notice that the analysis is developed under the desired expression, that is we only focus here on the 7th frame. Also, to highlight possible differences we work on the eyebrows, eyes and mouth regions, separately. For these subregions, the baseline chosen for the Bookstein coordinates are fixed at the following points: landmarks 1 and 6 for the eyebrows, the external corners (i.e. landmarks 11 and 19) for the eyes and the left and right corners for the mouth (i.e. landmarks 27 and 31).

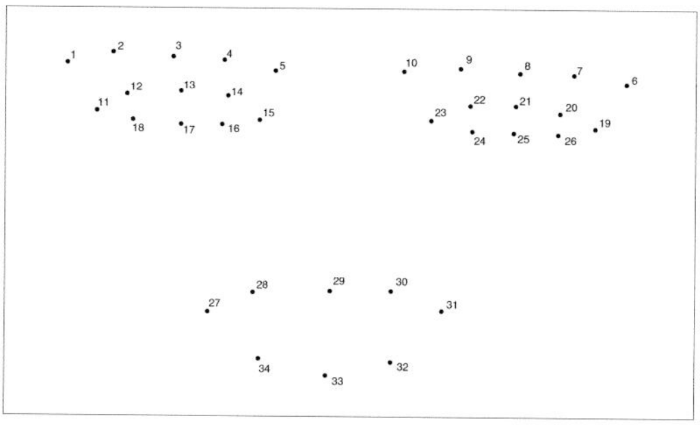

Fig. 2.1 Facial landmark configuration. The facial expression is summarized by 34 landmarks. The numbering scheme is consistent across all the frames and subject-configurations

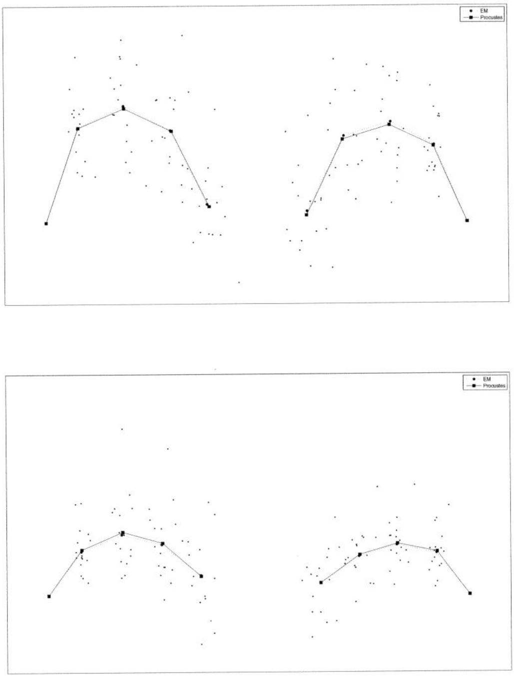

Fig. 2.2 Estimated mean paths of the eyebrows for *happiness* (*top*) and *surprise* (*bottom*) expressions using EM (*dashed line*) and Procrustes (*continuous line*) procedures

Results for the estimates of the mean paths for the complex case are shown in Figs. 2.2, 2.3 and 2.4. The estimated means under isotropy are very similar and are not shown here. The figures also show the GPA (Generalised Procrustes) estimate of the mean paths which appear very similar to the EM estimate.

In order to evaluate statistical differences between facial expressions we consider the values of the log-likelihoods for *happiness* and *surprise*. For the first expression the EM algorithm gives the following values of the log-likelihood: 625.33 (eyebrows), 1098.80 (eyelids) and 362.52 (mouth) for the isotropic case, and 783.22, 1548.70 and 468.98, for the complex covariance structure. Similarly, log-likelihood values for *surprise* are: 671.37, 1142.90 and 329.53 for the isotropic covariances, and 821.87, 1478.00 and 430.93 for the complex case.

Given the values of the likelihoods, differences between the expressions can be tested through the Generalised likelihood ratio test. In fact, if we are interested to test $H_0 : \Theta \in \Omega_0$ versus $H_1 : \Theta \in \Omega$, where $\Omega_0 \subset \Omega$, then for large samples and under regularity conditions, the likelihood ratio test is defined to be

$$-2\log\Lambda = 2\left(\sup_{H_1}\log L(\Theta) - \sup_{H_0}\log L(\Theta)\right)$$

which rejects H_0 for 'large' values.

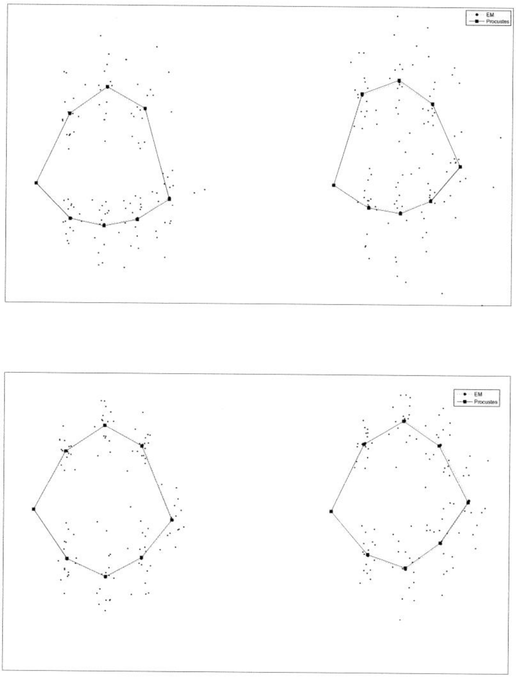

Fig. 2.3 Estimated mean paths of the eyelids for *happiness* (*top*) and *surprise* (*bottom*) expressions using EM (*dashed line*) and Procrustes (*continuous line*) procedures

For large samples and under H_0, we thus have the result that $-2\log \Delta$ has an approximate χ_r^2 distribution. The number of degrees of freedom, r, is equal to the number of free parameters under H_1 minus the number of free parameters under H_0 (i.e. $r = \dim(\Omega) - \dim(\Omega_0)$). This result allows us to use the χ_r^2 tables for finding rejection regions with a fixed value of α.

Hence, if we want to test whether the means of the two expressions differ from each other only by some rotation we consider the hypothesis test

$$\begin{cases} H_0 : \boldsymbol{\mu}_z^h = \boldsymbol{\mu}_z^s \mod (rot) \\ H_1 : \boldsymbol{\mu}_z^h \neq \boldsymbol{\mu}_z^s \mod (rot) \quad \boldsymbol{\Sigma}_z^h = \boldsymbol{\Sigma}_z^s. \end{cases}$$

We also assume that the preforms of *happiness* and *surprise* configurations are from complex normal distributions, that is $\mathscr{CN}_d\left(\boldsymbol{\mu}_z^h, \boldsymbol{\Sigma}_z^h\right)$ and $\mathscr{CN}_d\left(\boldsymbol{\mu}_z^s, \boldsymbol{\Sigma}_z^s\right)$, where $d = 9, 15, 7$ are the pre-form dimensions for the eyebrows, eyelids and mouth regions. The degrees of freedom for the eyebrows, eyelids and mouth are computed as $(2 \cdot 10) - 2 - 1 = 17$, $(2 \cdot 16) - 2 - 1 = 29$ and $(2 \cdot 8) - 2 - 1 = 13$ respectively. Also, by running the EM for the pooled samples, the log-likelihood value at the MLE estimates is 1523.90, 2768.71 and 830.76 for the eyebrows, eyelids and mouth, respectively. The likelihood values for the alternative hypothesis can be obtained by running the EM separately for each group while keeping the entries of $\boldsymbol{\Sigma}_z^h = \boldsymbol{\Sigma}_z^s = \boldsymbol{\Sigma}_z$. We obtain the maximum log-likelihood values 750.32 (eyebrows), 1432.12 (eyelids) and 445.67 (mouth) for the group of happiness, while

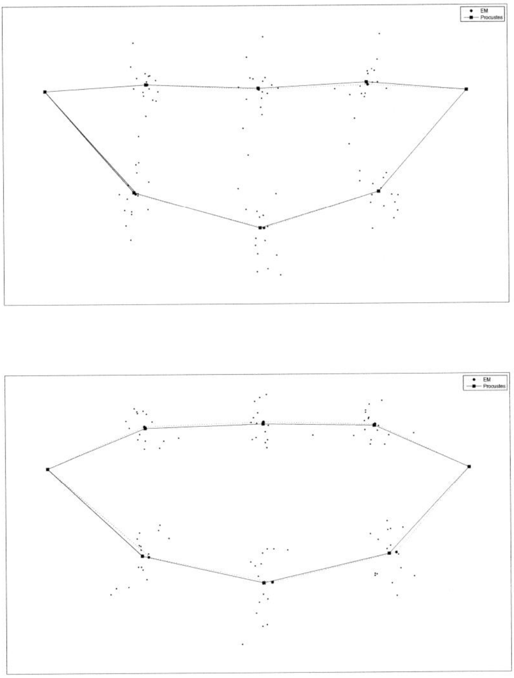

Fig. 2.4 Estimated mean paths of the mouth for *happiness* (*top*) and *surprise* (*bottom*) expressions using EM (*dashed line*) and Procrustes (*continuous line*) procedures

for the surprise we have 793.29 (eyebrows), 1362.80 (eyelids) and 410.09 (mouth). Hence, $-2\log\Delta$ for the eyebrows is distributed as χ^2_{17}, for the eyelids we have a χ^2_{29} and for the mouth the resulting distribution is a χ^2_{13}, which under H_0 are 39.42, 52.4 and 50, respectively. Since $P(\chi^2_{17} > 39.42)$, $P(\chi^2_{29} > 52.40)$ and $P(\chi^2_{13} > 50.00)$ are almost zero, there is a strong evidence that modulo rotations, $\boldsymbol{\mu}^h_z$ and $\boldsymbol{\mu}^s_z$ are different in all the constituent parts.

References

Adali T, Schreier PJ, Scharf LL (2011) Complex-valued signal processing: the proper way to deal with impropriety. IEEE Trans Signal Process 59(11):5101–5125

Dempster A, Laird N, Rubin D (1977) Maximum likelihood from incomplete data via the em algorithm. J R Stat Soc Ser B 34(1):1–38

Dryden IL, Mardia KV (1991) General shape distributions in the plane. Adv Appl Probab 23:259–276

Dryden IL, Mardia KV (1998) Statistical shape analysis. Wiley, London

Ekman P, Friesen WV (1971) Constants across culture in the face and emotion. J Pers Soc Psychol 17:124–129

Gradshteyn I, Ryzhik I (1980) Table of integrals, series, and products. Academic, New York

Kume A, Welling M (2010) Maximum likelihood estimation for the offset-normal shape distributions using em. J Comput Graph Stat 19:702–723

McLachlan G, Krishnan T (1997) The EM algorithm and extensions. Wiley, New York

Neeser F, Massey J (1993) Proper complex random processes with applications to information theory. IEEE Trans Inf Theory 39(4):1293–1302
Picinbono B (1996) Second-order complex random vectors and normal distributions. IEEE Trans Signal Process 44(10):2637–2640
Schreier P, Scharf L (2003) Second-order analysis of improper complex random vectors and processes. IEEE Trans Signal Process 51(3):714–725
Wallhoff F (2006) Facial expressions and emotion database. http://www.mmk.ei.tum.de/~waf/fgnet/feedtum.html

Chapter 3
Dynamic Shape Analysis Through the Offset-Normal Distribution

Abstract Statistical analysis of dynamic shapes is a problem with significant challenges due to the difficulty in providing a description of the shape changes over time, across subjects and over groups of subjects.

Recent attempts to study the shape change in time are based on the Procrustes tangent coordinates or spherical splines in Kendall shape spaces (Kent et al. (2001) Functional models of growth for landmark data. In: Proceedings in functional and spatial data analysis. University Press, Leeds, pp 109–115; Kume et al. Biometrika 94:513–528, 2007; Fishbaugh et al. (2012) Analysis of longitudinal shape variability via subject specific growth modeling. In: Ayache N, Delingette H, Golland P, Mori K (eds) Medical image computing and computer-assisted intervention – MICCAI 2012. Lecture notes in computer science, vol 7510. Springer, Berlin, Heidelberg, pp 731–738; Hinkle et al. (2012) International anthropometric study of facial morphology in various ethnic groups/races. In: Computer Vision - ECCV. Lecture Notes in Computer Science, vol 7574. pp 1–14; Fontanella et al. (2013) A functional spatio-temporal model for geometric shape analysis. In: Torelli N, Pesarin F, Bar-Hen A (eds) Advances in theoretical and applied statistics. Springer, Berlin, pp 75–86).

This chapter deals with the statistical analysis of a temporal sequence of landmark data using the exact distribution theory for the shape of planar correlated Gaussian configurations. Specifically, we extend the theory introduced in the second chapter to a dynamic framework and discuss the use of the offset-normal distribution for the description of time-varying shapes.

Modeling the temporal correlation structure of the dynamic process is a complex task, in general. For two time points, Mardia and Walder (Biometrika 81:185–196, 1994) have shown that the density function of the offset-normal distribution has a rather complicated form and have discussed the difficulty of extending their results to $t > 2$. In the final part of the chapter we show that, in principle, it is possible to calculate the closed form expression of the offset-normal distribution for a general t, though its calculation can be computationally expensive.

This chapter is organized as follows. In Sect. 3.1 we describe the offset-normal shape distribution in a dynamic context. In Sect. 3.2 we introduce the EM algorithm for general spatio-temporal covariance matrices while Sect. 3.3 describes the necessary adjustments of the general update rules under separability assumptions of the spatio-temporal covariance structure. A discussion of the computational difficulties

C. Brombin et al., *Parametric and Nonparametric Inference for Statistical Dynamic Shape Analysis with Applications*, SpringerBriefs in Statistics,
DOI 10.1007/978-3-319-26311-3_3

concerning the performance of the algorithm is also provided. Following Kume and Welling (J Comput Graph Stat 19:702–723, 2010), in Sect. 3.4 we discuss the case in which the temporal dynamics of the shape variables are only modeled through a polynomial regression which captures the large-scale temporal variability of the process. The fit of this regression model to *happiness* and *surprise* data are shown in Sect. 3.4.1. For the same expressions, we also consider the problem of matching symmetry and provide some comments in Sect. 3.5. Finally, Sect. 3.6 concludes the chapter by discussing the use of mixture models for classification purposes in a dynamic setting.

Keywords Statistical shape analysis • EM Algorithm • Time-varying shapes • Polynomial shape regression • Moments of quadratic forms • Bilateral Symmetry • Mixture models • Facial expressions

3.1 The Offset-Normal Distribution in a Dynamic Setting

Suppose that for planar shapes, landmark data are available at times $t = 1, \ldots, T$. Since at time t, the landmark Cartesian coordinates are organized in the $(K \times 2)$ configuration matrix $\mathbf{X}_t^\dagger = (\mathbf{x}_t^\dagger \; \mathbf{y}_t^\dagger)$, the temporal sequence of the T configuration matrices can be arranged in a $K \times 2T$ data matrix, $\mathbf{X}^\dagger = \left[\mathbf{X}_1^\dagger \; \mathbf{X}_2^\dagger \; \ldots \; \mathbf{X}_T^\dagger\right]$.

The pre-form coordinates matrix is obtained by removing the translation effect at each time t. This is achieved by pre-multiplying each configuration matrix $\mathbf{X}_t^\dagger$, for $t = 1, \ldots, T$, with the $(K-1)\times K$ matrix $\mathbf{L}$. Therefore, the temporal sequence of pre-form of configurations is $\mathbf{X} = \mathbf{L}\mathbf{X}^\dagger = \left(\mathbf{L}\mathbf{X}_1^\dagger \; \mathbf{L}\mathbf{X}_2^\dagger \; \ldots \; \mathbf{L}\mathbf{X}_T^\dagger\right) = (\mathbf{X}_1 \; \mathbf{X}_2 \; \ldots \; \mathbf{X}_T)$.

In order to work with shape variables, it is also necessary to remove the information about scaling and rotation. By considering Bookstein coordinates, at each time t we work with the transformation $\mathbf{U}_t \rightarrow \varphi_t \mathbf{X}_t \mathbf{R}_t$, where the scaling factor and the rotation matrix are given by

$$\varphi_t = (x_{2,t}^2 + y_{2,t}^2)^{-1}, \quad \mathbf{R}_t = \begin{pmatrix} x_{2_t} & -y_{2_t} \\ y_{2_t} & x_{2_t} \end{pmatrix} \qquad t = 1, \ldots, T$$

with $x_{2,t}^2 + y_{2,t}^2 = |\mathbf{R}_t|$. The transformation $\mathbf{X} \rightarrow \mathbf{U}$, thus gives the full set of shape variables

$$(\mathbf{U}_1 \; \mathbf{U}_2 \; \ldots \; \mathbf{U}_T) = \left(\mathbf{X}_1\mathbf{R}_1^* \; \mathbf{X}_2\mathbf{R}_2^* \; \ldots \; \mathbf{X}_T\mathbf{R}_T^*\right)$$

or, in matrix formulation, $\mathbf{U} = \mathbf{X}\mathbf{R}^*$, where $\mathbf{R}^* = diag\left(\mathbf{R}_1^*, \mathbf{R}_2^*, \ldots, \mathbf{R}_T^*\right)$ is a $2T \times 2T$ block-diagonal matrix and $\mathbf{R}_t^* = \varphi_t \mathbf{R}_t$.

In complex notation, the $K \times T$ matrix of the sequence of configurations is denoted as $\mathbf{Z}^{\dagger} = \left(\mathbf{z}_1^{\dagger}\ \mathbf{z}_2^{\dagger}\ \ldots\ \mathbf{z}_T^{\dagger}\right) = \left(\mathbf{x}_1^{\dagger} + i\mathbf{y}_1^{\dagger}\ \mathbf{x}_2^{\dagger} + i\mathbf{y}_2^{\dagger}\ \ldots\ \mathbf{x}_T^{\dagger} + i\mathbf{y}_T^{\dagger}\right)$, and the coordinates in the pre-form space are given by $\mathbf{Z} = \mathbf{L}\mathbf{Z}^{\dagger}$.

Then, at time t, scaling and rotation effects are filtered out by computing the ratio $\boldsymbol{\xi}_t = \mathbf{z}_t / z_{2,t}$, where $\boldsymbol{\xi}_t = \boldsymbol{u}_t + i\boldsymbol{v}_t$. In matrix form, the complete $(K-1) \times T$ matrix is thus obtained as $\boldsymbol{\xi} = (\boldsymbol{\xi}_1\ \boldsymbol{\xi}_2\ \ldots\ \boldsymbol{\xi}_T) = (\mathbf{z}_1\ \mathbf{z}_2\ \ldots\ \mathbf{z}_T)\, diag\left(\frac{1}{z_{2,1}}, \frac{1}{z_{2,2}}, \ldots, \frac{1}{z_{2,T}}\right)$.

3.1.1 The Probability Density Function

Assume that, in the configuration space, for a temporal sequence of landmark coordinates, the $(2KT \times 1)$ vector $vec(\mathbf{X}^{\dagger}) = \left(\mathbf{x}_1^{\dagger'}\ \mathbf{y}_1^{\dagger'} \ldots \mathbf{x}_T^{\dagger'}\ \mathbf{y}_T^{\dagger'}\right)'$ is distributed as $\mathcal{N}_{2KT}\left(vec(\boldsymbol{\mu}^{\dagger}), \boldsymbol{\Sigma}^{\dagger}\right)$, where $vec(\boldsymbol{\mu}^{\dagger}) = \left(vec(\boldsymbol{\mu}_1^{\dagger'})\ \ldots\ vec(\boldsymbol{\mu}_T^{\dagger})'\right)$ and

$$\boldsymbol{\Sigma}^{\dagger} = \begin{pmatrix} \boldsymbol{\Sigma}^{\dagger}_{x_1x_1} & \boldsymbol{\Sigma}^{\dagger}_{x_1y_1} & \ldots & \boldsymbol{\Sigma}^{\dagger}_{x_1x_T} & \boldsymbol{\Sigma}^{\dagger}_{x_1y_T} \\ \boldsymbol{\Sigma}^{\dagger}_{y_1x_1} & \boldsymbol{\Sigma}^{\dagger}_{y_1y_1} & \ldots & \boldsymbol{\Sigma}^{\dagger}_{y_1x_T} & \boldsymbol{\Sigma}^{\dagger}_{y_1y_T} \\ \vdots & \vdots & \vdots & \vdots & \vdots \\ \boldsymbol{\Sigma}^{\dagger}_{x_Tx_1} & \boldsymbol{\Sigma}^{\dagger}_{x_Ty_1} & \ldots & \boldsymbol{\Sigma}^{\dagger}_{x_Tx_T} & \boldsymbol{\Sigma}^{\dagger}_{x_Ty_T} \\ \boldsymbol{\Sigma}^{\dagger}_{y_Tx_1} & \boldsymbol{\Sigma}^{\dagger}_{y_Ty_1} & \ldots & \boldsymbol{\Sigma}^{\dagger}_{y_Tx_T} & \boldsymbol{\Sigma}^{\dagger}_{y_Ty_T} \end{pmatrix} = \begin{pmatrix} \boldsymbol{\Sigma}^{\dagger}_{1} & \ldots & \boldsymbol{\Sigma}^{\dagger}_{1T} \\ \vdots & \vdots & \vdots \\ \boldsymbol{\Sigma}^{\dagger}_{T1} & \ldots & \boldsymbol{\Sigma}^{\dagger}_{T} \end{pmatrix}$$

where $\boldsymbol{\Sigma}^{\dagger}_{x_tx_t}$, $\boldsymbol{\Sigma}^{\dagger}_{y_ty_t}$ and $\boldsymbol{\Sigma}^{\dagger}_{x_ty_t}$ are $K \times K$ landmark covariance matrices computed at a specific time point, t, while $\boldsymbol{\Sigma}^{\dagger}_{x_tx_{t'}}$, $\boldsymbol{\Sigma}^{\dagger}_{y_ty_{t'}}$ and $\boldsymbol{\Sigma}^{\dagger}_{x_ty_{t'}}$ are $(K \times K)$ landmark covariance matrices at two different times, t and t'. We have $\boldsymbol{\Sigma}^{\dagger}_{x_ty_t} = (\boldsymbol{\Sigma}^{\dagger}_{y_tx_t})'$, $\boldsymbol{\Sigma}^{\dagger}_{x_ty_{t'}} = (\boldsymbol{\Sigma}^{\dagger}_{y_{t'}x_t})'$ and $\boldsymbol{\Sigma}^{\dagger}_{x_ty_{t'}} \neq \boldsymbol{\Sigma}^{\dagger}_{x_{t'}y_t}$. In addition, $\boldsymbol{\Sigma}^{\dagger}_{t}$ and $\boldsymbol{\Sigma}^{\dagger}_{tt'}$ are the autocovariance and cross-covariance matrices for configurations $\mathbf{X}_t$ and $\mathbf{X}_{t'}$, respectively. Accordingly, in the pre-form space, we have $vec(\mathbf{X}) = vec(\mathbf{L}\mathbf{X}^{\dagger}) \sim \mathcal{N}_{2(K-1)T}\left(vec(\boldsymbol{\mu}), \boldsymbol{\Sigma}\right)$, so that

$$f\left(vec(\mathbf{X})|\boldsymbol{\mu}, \boldsymbol{\Sigma}\right) = \frac{1}{(2\pi)^{(K-1)T}|\boldsymbol{\Sigma}|^{\frac{1}{2}}} exp\left\{-\frac{1}{2}\left[vec(\mathbf{X}) - vec(\boldsymbol{\mu})\right]' \right.$$
$$\left. \boldsymbol{\Sigma}^{-1}\left[vec(\mathbf{X}) - vec(\boldsymbol{\mu})\right]\right\} \tag{3.1}$$

where $vec(\boldsymbol{\mu}) = vec(\mathbf{L}\boldsymbol{\mu}^{\dagger}) = (\mathbf{I}_{2T} \otimes \mathbf{L})vec(\boldsymbol{\mu}^{\dagger})$ and $\boldsymbol{\Sigma} = (\mathbf{I}_{2T} \otimes \mathbf{L})\boldsymbol{\Sigma}^{\dagger}(\mathbf{I}_{2T} \otimes \mathbf{L}')$.

As discussed in Chap. 2, the distribution of the shape variables, $\mathbf{u} = \{u_{k,t}, v_{k,t}\}_{k=3:K;t=1:T}$, can be obtained by integrating out the vector, $\mathbf{h} = \left(\mathbf{h}_1' \ldots \mathbf{h}_T'\right) = (x_{2,1}, y_{2,1}, \ldots, x_{2,T}, y_{2,T})'$, where the vector $\mathbf{h}_t = (x_{2,t}, y_{2,t})'$ represents the rotation and scaling information for the pre-form $\mathbf{X}_t$.

Considering T configurations, we still consider the transformation $vec(\mathbf{X}) = \mathbf{W}\mathbf{h}$ but now $\mathbf{W} = diag\,(\mathbf{W}_1, \ldots, \mathbf{W}_T)$ is a $2T(K-1) \times 2T$ block-diagonal matrix with

$$\mathbf{W}_t = \begin{pmatrix} 1 & u_{3,t} & \dots & u_{K,t} & 0 & v_{3,t} & \dots & v_{K,t} \\ 0 & -v_{3,t} & \dots & -v_{K,t} & 1 & u_{3,t} & \dots & u_{K,t} \end{pmatrix}', \quad t = 1, \dots, T.$$

Since the Jacobian of the transformation $\mathbf{X} \to (\mathbf{h}, \mathbf{u})$ is given by $\prod_{t=1}^{T} \|\mathbf{h}_t\|^{2(K-2)}$, the joint distribution of $(\mathbf{h}, \mathbf{u})$ can be written as

$$f(\mathbf{h}, \mathbf{u}|\boldsymbol{\mu}, \boldsymbol{\Sigma}) = \frac{1}{(2\pi)^{(K-1)T}|\boldsymbol{\Sigma}|^{\frac{1}{2}}} exp\left\{-\frac{1}{2}[\mathbf{Wh} - vec(\boldsymbol{\mu})]'\boldsymbol{\Sigma}^{-1}[\mathbf{Wh} - vec(\boldsymbol{\mu})]\right\} \times \prod_{t=1}^{T} \|\mathbf{h}_t\|^{2(K-2)}. \tag{3.2}$$

Then, following Sect. 2.4, by defining the $2T \times 2T$ matrix, $\boldsymbol{\Gamma} = (\mathbf{W}'\boldsymbol{\Sigma}^{-1}\mathbf{W})^{-1}$, and the $2T$-dimensional vector, $\boldsymbol{\eta} = \boldsymbol{\Gamma}\mathbf{W}'\boldsymbol{\Sigma}^{-1}vec(\boldsymbol{\mu})$, the joint distribution (3.2) can be rewritten as

$$f(\mathbf{h}, \mathbf{u}|\boldsymbol{\mu}, \boldsymbol{\Sigma}) = \frac{exp(-g/2)}{(2\pi)^{(K-1)T}|\boldsymbol{\Sigma}|^{\frac{1}{2}}} \cdot exp\left\{-\frac{(\mathbf{h}-\boldsymbol{\eta})'\boldsymbol{\Gamma}^{-1}(\mathbf{h}-\boldsymbol{\eta})}{2}\right\} \prod_{t=1}^{T} \|\mathbf{h}_t\|^{2(K-2)} \tag{3.3}$$

where $g = vec(\boldsymbol{\mu})'\boldsymbol{\Sigma}^{-1}vec(\boldsymbol{\mu}) - \boldsymbol{\eta}'\boldsymbol{\Gamma}^{-1}\boldsymbol{\eta}$.

Finally, the off-set normal shape density function is obtained by integrating out $\mathbf{h}$

$$\begin{aligned} f(\mathbf{u}|\boldsymbol{\mu}, \boldsymbol{\Sigma}) &= \frac{exp(-g/2)|\boldsymbol{\Gamma}|^{\frac{1}{2}}}{(2\pi)^{(K-2)T}|\boldsymbol{\Sigma}|^{\frac{1}{2}}} \int \prod_{t=1}^{T} \|\mathbf{h}_t\|^{2(K-2)} f_{\mathcal{N}_{2T}}(\mathbf{h}|\boldsymbol{\eta}, \boldsymbol{\Gamma}) d\mathbf{h} \\ &= \frac{exp(-g/2)|\boldsymbol{\Gamma}|^{\frac{1}{2}}}{(2\pi)^{(K-2)T}|\boldsymbol{\Sigma}|^{\frac{1}{2}}} \int \prod_{t=1}^{T} (\mathbf{h}'\mathbf{A}_t\mathbf{h})^{K-2} f_{\mathcal{N}_{2T}}(\mathbf{h}|\boldsymbol{\eta}, \boldsymbol{\Gamma}) d\mathbf{h} \\ &= \frac{exp(-g/2)|\boldsymbol{\Gamma}|^{\frac{1}{2}}}{(2\pi)^{(K-2)T}|\boldsymbol{\Sigma}|^{\frac{1}{2}}} E\left[\prod_{t=1}^{T} (\mathbf{h}'\mathbf{A}_t\mathbf{h})^{K-2}\right] \end{aligned} \tag{3.4}$$

where $\mathbf{A}_t = diag(\mathbf{0}_{2t-2}, \mathbf{I}_2, \mathbf{0}_{2T-2t})$, with $\mathbf{0}_t$ a $t \times t$ null matrix and $\mathbf{I}_t$ the identity matrix of dimension t. We notice that $E\left[\prod_{t=1}^{T} (\mathbf{h}'\mathbf{A}_t\mathbf{h})^{k-2}\right]$ denotes the moments of a product of quadratic forms of noncentral normal random variables, $\mathbf{h} \sim \mathcal{N}_{2T}(\boldsymbol{\eta}, \boldsymbol{\Gamma})$.

For the computation of these moments, we can use the following expansion proposed by Kan (2008)

$$E\left[\prod_{t=1}^{T} (\mathbf{h}'\mathbf{A}_t\mathbf{h})^{K-2}\right] = \frac{1}{s!} \sum_{v_1=0}^{s_1} \cdots \sum_{v_T=0}^{s_T} (-1)^{\sum_{t=1}^{T} v_t} \binom{s_1}{v_1} \cdots \binom{s_T}{v_T} Q_s(\mathbf{B}_v) \tag{3.5}$$

where $Q_s(\mathbf{B}_v) = E[(\mathbf{h}'\mathbf{B}_v\mathbf{h})^s]$, $\mathbf{B}_v = \sum_{t=1}^{T}\left(\frac{s_t}{2} - v_t\right)\mathbf{A}_t$, $s = T(K-2)$ and $s_t = K-2$, $\forall t$. A solution for $Q_s(\mathbf{B}_v)$, can be found by following Lemma 2 of Magnus (1986) which suggests the following expression

$$E\left[(\mathbf{h}'\mathbf{B}_v\mathbf{h})^s\right] = 2^s s! \sum_{\delta}\prod_{j=1}^{s} \frac{\left(tr(\mathbf{B}_v\boldsymbol{\Gamma})^j + j\boldsymbol{\eta}'(\mathbf{B}_v\boldsymbol{\Gamma})^{j-1}\mathbf{B}_v\boldsymbol{\eta}\right)^{\delta_j}}{\delta_j!(2j)^{\delta_j}} \tag{3.6}$$

with the summation over all s-vector $\boldsymbol{\delta} = (\delta_1, \ldots, \delta_s)$, whose elements are nonnegative integers satisfying $\sum_{j=1}^{s} j\delta_j = s$. However, this expression requires an algorithm to enumerate all the partitions of the integer s.

An expression for $Q_s(\mathbf{B}_v)$ that is computationally more efficient than Eq. (3.6) is based on the recursive relation between moments and cumulants (see, for example, Mathai and Provost 1992, Eq.3.2b.8) and is given by

$$E\left[(\mathbf{h}'\mathbf{B}_v\mathbf{h})^s\right] = s!2^s d_s(\mathbf{B}_v) \tag{3.7}$$

where

$$d_s(\mathbf{B}_v) = \frac{1}{2s}\sum_{j=1}^{s}\left[tr(\mathbf{B}_v\boldsymbol{\Gamma})^j + j\boldsymbol{\eta}'(\mathbf{B}_v\boldsymbol{\Gamma})^{j-1}\mathbf{B}_v\boldsymbol{\eta}\right] d_{s-1}(\mathbf{B}_v), \quad d_0(\mathbf{B}_v) = 1.$$

As noticed by Kan (2008), although Eq. (3.7) does not provide an explicit expression for $Q_s(\mathbf{B}_v)$, it is easier to program than (3.6) and it also takes much less time to compute.

When $s = T(K-2)$ is large, in order to compute $tr(\mathbf{B}_v\boldsymbol{\Gamma})^j + j\boldsymbol{\eta}'(\mathbf{B}_v\boldsymbol{\Gamma})^{j-1}\mathbf{B}_v\boldsymbol{\eta}$, Kan (2008) suggests to perform an eigenvalue decomposition. If $\boldsymbol{\Gamma}$ is positive definite, we can consider the Cholesky decomposition $\boldsymbol{\Gamma} = \mathbf{L}_\Gamma\mathbf{L}_\Gamma'$ and the eigenvalue decomposition $\mathbf{L}_\Gamma'\mathbf{B}_v\mathbf{L}_\Gamma = \mathbf{P}_v\boldsymbol{\Lambda}_v\mathbf{P}_v'$, such that assuming there are $c \leq 2T$ nonzero eigenvalues we have that

$$tr(\mathbf{B}_v\boldsymbol{\Gamma})^j + j\boldsymbol{\eta}'(\mathbf{B}_v\boldsymbol{\Gamma})^{j-1}\mathbf{B}_v\boldsymbol{\eta} = tr(\boldsymbol{\Lambda}_v)^j + j\tilde{\boldsymbol{\eta}}'(\boldsymbol{\Lambda}_v)^j\tilde{\boldsymbol{\eta}} = \sum_{t=1}^{c}(1 + j\tilde{\eta}_t^2)\lambda_t^j$$

where $\tilde{\boldsymbol{\eta}} = \mathbf{P}_v'\mathbf{L}_\Gamma^{-1}\boldsymbol{\eta}$.

3.2 EM Algorithm for Estimating μ and Σ

In this section we discuss the EM algorithm for a temporal sequence of shapes.

Assume that $\mathscr{X} = \{\mathbf{X}^{(n)}\}_{n=1:N}$ and $\mathscr{U} = \{\mathbf{u}^{(n)}\}_{n=1:N}$ denote the full data and the observed (shape) data, respectively, for a random sample of N sequences of landmark configurations. By following the same lines of Chap. 2, the maximum of the conditional log-likelihood (*M-step*), $\mathcal{Q}_{\boldsymbol{\mu}^{(r)},\boldsymbol{\Sigma}^{(r)}}(\boldsymbol{\mu}, \boldsymbol{\Sigma})$, is achieved at

$$vec(\boldsymbol{\mu}^{(r+1)}) = \frac{1}{N}\sum_{n=1}^{N}\int vec(\mathbf{X}^{(n)})dF(\mathbf{X}^{(n)}|\mathbf{u}^{(n)}, \boldsymbol{\mu}^{(r)}, \boldsymbol{\Sigma}^{(r)})$$

$$\boldsymbol{\Sigma}^{(r+1)} = \frac{1}{N}\sum_{n=1}^{N}\int vec\left(\mathbf{X}^{(n)}\right) vec\left(\mathbf{X}^{(n)}\right)' dF(\mathbf{X}^{(n)}|\mathbf{u}^{(n)}, \boldsymbol{\mu}^{(r)}, \boldsymbol{\Sigma}^{(r)}) - vec(\boldsymbol{\mu}_{r+1})vec(\boldsymbol{\mu}_{r+1})',$$

and the expectations which establish the update rules for the parameters (*E step*) are given by

$$\int vec(\mathbf{X})dF(\mathbf{X}|\mathbf{u}, \boldsymbol{\mu}, \boldsymbol{\Sigma}) = \mathbf{W}\frac{\int_{\Re^{2T}} \mathbf{h}f(\mathbf{h}, \mathbf{u}|\boldsymbol{\mu}, \boldsymbol{\Sigma})d\mathbf{h}}{f(\mathbf{u}|\boldsymbol{\mu}^{(r)}, \boldsymbol{\Sigma}^{(r)})} \tag{3.8}$$

$$\int vec(\mathbf{X})vec(\mathbf{X})'dF(\mathbf{X}|\mathbf{u}, \boldsymbol{\mu}, \boldsymbol{\Sigma}) = \mathbf{W}\frac{\int_{\Re^{2T}} \mathbf{h}\mathbf{h}'f(\mathbf{h}, \mathbf{u}|\boldsymbol{\mu}, \boldsymbol{\Sigma})d\mathbf{h}}{f(\mathbf{u}|\boldsymbol{\mu}^{(r)}, \boldsymbol{\Sigma}^{(r)})}\mathbf{W}'. \tag{3.9}$$

Given $\mathbf{h} \sim \mathcal{N}_{2T}(\boldsymbol{\eta}, \boldsymbol{\Gamma})$, in order to develop the EM algorithm, it is convenient to write the quadratic form $Q_s(\mathbf{B}_v)$ of the induced pdf in Eq. (3.4) as (Mathai and Provost 1992, p. 28):

$$Q_s(\mathbf{B}_v) = E\left[(\mathbf{h}'\mathbf{B}_v\mathbf{h})^s\right] = E\left\{\left[(\mathbf{l} + \mathbf{L}_{\Gamma}^{-1}\boldsymbol{\eta})'\mathbf{L}_{\Gamma}'\mathbf{B}_v\mathbf{L}_{\Gamma}(\mathbf{l} + \mathbf{L}_{\Gamma}^{-1}\boldsymbol{\eta})\right]^s\right\} \tag{3.10}$$

where $\mathbf{L}_{\Gamma}$ is a lower triangular matrix with $\boldsymbol{\Gamma} = \mathbf{L}_{\Gamma}\mathbf{L}_{\Gamma}'$, $\mathbf{l} = \mathbf{L}_{\Gamma}^{-1}(\mathbf{h} - \boldsymbol{\eta})$, $E[\mathbf{l}] = \mathbf{0}$ and $Cov(\mathbf{l}) = \mathbf{I}_{2t}$.

Considering the eigen-decomposition $\mathbf{L}_{\Gamma}'\mathbf{B}_v\mathbf{L}_{\Gamma} = \mathbf{P}_v\boldsymbol{\Lambda}_v\mathbf{P}_v'$, where $\mathbf{P}_v$ is the matrix of eigenvectors and $\boldsymbol{\Lambda}_v$ the diagonal matrix of the corresponding eigenvalues, it follows

$$E\left[(\mathbf{h}'\mathbf{B}_v\mathbf{h})^s\right] = E\left\{[(\mathbf{l}_v + \boldsymbol{\zeta}_v)'\boldsymbol{\Lambda}_v(\mathbf{l}_v + \boldsymbol{\zeta}_v)]^s\right\} = E\left\{\left[\sum_{t=1}^{2T}\lambda_{v_t}(l_{v_t} + \zeta_{v_t})^2\right]^s\right\}$$
$$= \sum_{t=1}^{2T}\lambda_{v_t}\nu_{v_t}^2 \tag{3.11}$$

where, $\nu_{v_t} = l_{v_t} + \zeta_{v_t}$, with l_{v_t} and ζ_{v_t} being the elements of the $2T$–dimensional vectors, $\mathbf{l}_v = \mathbf{P}_v'\mathbf{l}$ and $\boldsymbol{\zeta}_v = \mathbf{P}_v'\mathbf{L}_{\Gamma}^{-1}\boldsymbol{\eta}$.

Therefore, the marginal (off-set normal shape) pdf can be written as

$$f(\mathbf{u}|\boldsymbol{\mu}, \boldsymbol{\Sigma}) = \frac{exp(-g/2)|\boldsymbol{\Gamma}|^{\frac{1}{2}}}{(2\pi)^{(K-2)T}|\boldsymbol{\Sigma}|^{\frac{1}{2}}} \frac{1}{s!} \sum_{v_1=0}^{s_1} \cdots \sum_{v_T=0}^{s_T} (-1)^{\sum_{t=1}^{T} v_t} \binom{s_1}{v_1} \cdots \binom{s_T}{v_T}$$
$$E\left[\left(\sum_{t=1}^{2T} \lambda_{v_t} \nu_{v_t}^2\right)^s\right]. \tag{3.12}$$

The computation of the integer moments $E\left[\left(\sum_{t=1}^{2T} \lambda_{v_t} \nu_{v_t}^2\right)^s\right]$ can be obtained as (Mathai and Provost 1992, p. 49)

$$E\left[\left(\sum_{t=1}^{2T} \lambda_{v_t} \nu_{v_t}^2\right)^s\right] = s! \sum_{p_1+} \cdots \sum_{+p_{2T}=s} \frac{\lambda_{v_1}^{p_1} \cdots \lambda_{v_{2T}}^{p_{2T}}}{p_1! \cdots p_{2T}!} E[\nu_{v_1}^{2p_1} \cdots \nu_{v_{2T}}^{2p_{2T}}] \tag{3.13}$$

where the summations are over all the partitions $p_1 + p_2 + \cdots + p_{2T} = s$.

Since the ν_{v_t}'s are independent Gaussian random variables with mean ζ_{v_t} and unit variance (see Theorem 3.2b.1 Mathai and Provost 1992, p. 49), the marginal off-set normal pdf is finally given by

$$f(\mathbf{u}|\boldsymbol{\mu}, \boldsymbol{\Sigma}) = \frac{exp(-g/2)|\boldsymbol{\Gamma}|^{\frac{1}{2}}}{(2\pi)^{(K-2)T}|\boldsymbol{\Sigma}|^{\frac{1}{2}}} \sum_{v_1=0}^{s_1} \cdots \sum_{v_T=0}^{s_T} (-1)^{\sum_{t=1}^{T} v_t} \binom{s_1}{v_1} \cdots \binom{s_T}{v_T}$$
$$\sum_{p_1+} \cdots \sum_{+p_{2T}=s} \frac{\lambda_{v_1}^{p_1} \cdots \lambda_{v_{2T}}^{p_{2T}}}{p_1! \cdots p_{2T}!} E[\nu_{v_1}^{2p_1}|\zeta_{v_1}, 1] \cdots E[\nu_{v_{2T}}^{2p_{2T}}|\zeta_{v_{2T}}, 1]. \tag{3.14}$$

In Eq. (3.14), $E[\nu_{v_t}^{2p_t}] = E[\nu_{v_t}^{2p_t}|\zeta_{v_t}, 1]$ denotes the moments of the univariate Gaussian distribution with mean ζ_{v_t} and variance 1 and can be calculated as discussed in Chap. 2—see Eq. (2.6)

$$E[\nu_{v_t}^{2p_t}|\zeta_{v_t}, 1] = 2^{p_t} p_t! \, L_{p_t}^{-1/2}\left(-\frac{\zeta_{v_t}^2}{2}\right), \quad t = 1, \ldots, 2T$$

where $L_q^{\alpha}(\cdot)$ is the generalized Laguerre polynomial of order q.

The update rule for the mean in Eq. (3.8) also requires the solution of the integral

$$\int \mathbf{h} f(\mathbf{h}, \mathbf{u}|\boldsymbol{\mu}, \boldsymbol{\Sigma}) d\mathbf{h} = \frac{exp(-g/2)|\boldsymbol{\Gamma}|^{\frac{1}{2}}}{(2\pi)^{(k-2)T}|\boldsymbol{\Sigma}|^{\frac{1}{2}}} \int \mathbf{h} \prod_{t=1}^{T} (\mathbf{h}'\mathbf{A}_t\mathbf{h})^{k-2} f_{\mathcal{N}_{2T}}(\mathbf{h}|\boldsymbol{\eta}, \boldsymbol{\Gamma}) d\mathbf{h}.$$

Given $\mathbf{h} = \mathbf{L}_\Gamma \mathbf{P}_v \boldsymbol{\nu}_v$ and letting $M = \frac{exp(-g/2)|\boldsymbol{\Gamma}|^{\frac{1}{2}}}{(2\pi)^{(K-2)T}|\boldsymbol{\Sigma}|^{\frac{1}{2}}}$, it follows

$$\int \mathbf{h}f(\mathbf{h}, \mathbf{u}|\boldsymbol{\mu}, \boldsymbol{\Sigma})d\mathbf{h} = M \int \frac{1}{s!} \sum_{v_1=0}^{s_1} \cdots \sum_{v_T=0}^{s_T} (-1)^{\sum_{t=1}^T v_t}$$

$$\binom{s_1}{v_1} \cdots \binom{s_T}{v_T} (\mathbf{h}'\mathbf{B}_v\mathbf{h})^s \mathbf{h} f_{\mathcal{N}_{2T}}(\mathbf{h}|\boldsymbol{\eta}, \boldsymbol{\Gamma})d\mathbf{h}$$

$$= M\frac{1}{s!} \sum_{v_1=0}^{s_1} \cdots \sum_{v_T=0}^{s_T} (-1)^{\sum_{t=1}^T v_t} \binom{s_1}{v_1} \cdots \binom{s_T}{v_T}$$

$$\int (\boldsymbol{\nu}_v' \boldsymbol{\Lambda}_v \boldsymbol{\nu}_v)^s \mathbf{L}_\Gamma \mathbf{P}_v \boldsymbol{\nu}_v f_{\mathcal{N}_{2T}}(\boldsymbol{\nu}_v|\boldsymbol{\zeta}_v, \mathbf{I})d\boldsymbol{\nu}_v =$$

$$= M\mathbf{L}_\Gamma \sum_{v_1=0}^{s_1} \cdots \sum_{v_T=0}^{s_T} (-1)^{\sum_{t=1}^T v_t} \binom{s_1}{v_1} \cdots \binom{s_T}{v_T}$$

$$\mathbf{P}_v \sum_{p_1+} \cdots \sum_{+p_{2T}=s} \frac{\lambda_{v_1}^{p_1} \cdots \lambda_{v_{2T}}^{p_{2T}}}{p_1! \cdots p_{2T}!} E\left[(\nu_{v_1}^{2p_1} \cdots \nu_{v_{2T}}^{2p_{2T}})\boldsymbol{\nu}_v\right].$$

The jth entry of $E\left[(\nu_{v_1}^{2p_1} \cdots \nu_{v_{2T}}^{2p_{2T}})\boldsymbol{\nu}_v\right]$ is $E[\nu_{v_1}^{2p_1}|\zeta_{v_1}, 1] \cdots E[\nu_{v_j}^{2p_j+1}|\zeta_{v_j}, 1] \cdots E[\nu_{v_{2T}}^{2p_{2T}}|\zeta_{v_{2T}}, 1]$ where $E[\nu_{v_t}^{2p_t}|\zeta_{v_t}, 1] = 2^{p_t} p_t! \mathcal{L}_{p_t}^{-1/2}\left(-\frac{\zeta_{v_t}^2}{2}\right)$ and $E[\nu_{v_t}^{2p_t+1}|\zeta_{v_t}, 1] = \zeta_{v_t} 2^{p_t} p_t! \mathcal{L}_{p_t}^{1/2}\left(-\frac{\zeta_{v_t}^2}{2}\right)$.

In addition, the update rule for the covariance matrix in Eq. (3.9) requires the solution of the integral

$$\int \mathbf{h}\mathbf{h}'f(\mathbf{h}, \mathbf{u}|\boldsymbol{\mu}, \boldsymbol{\Sigma})d\mathbf{h} = \frac{exp(-g/2)|\boldsymbol{\Gamma}|^{\frac{1}{2}}}{(2\pi)^{(k-2)T}|\boldsymbol{\Sigma}|^{\frac{1}{2}}} \int \mathbf{h}\mathbf{h}' \prod_{t=1}^{T} (\mathbf{h}'\mathbf{A}_t\mathbf{h})^{k-2} f_{\mathcal{N}_{2T}}(\mathbf{h}|\boldsymbol{\eta}, \boldsymbol{\Gamma})d\mathbf{h}$$

where the expectation can be computed as

$$\int \mathbf{h}\mathbf{h}'f(\mathbf{h}, \mathbf{u}|\boldsymbol{\mu}, \boldsymbol{\Sigma})d\mathbf{h} = M \int \frac{1}{s!} \sum_{v_1=0}^{s_1} \cdots \sum_{v_T=0}^{s_T} (-1)^{\sum_{t=1}^T v_t}$$

$$\binom{s_1}{v_1} \cdots \binom{s_T}{v_T} (\mathbf{h}'\mathbf{B}_v\mathbf{h})^s \mathbf{h}\mathbf{h}' f_{\mathcal{N}_{2T}}(\mathbf{h}|\boldsymbol{\eta}, \boldsymbol{\Gamma})d\mathbf{h}$$

$$= M\mathbf{L}_\Gamma \sum_{v_1=0}^{s_1} \cdots \sum_{v_T=0}^{s_T} (-1)^{\sum_{t=1}^T v_t} \binom{s_1}{v_1} \cdots \binom{s_T}{v_T}$$

$$\mathbf{P}_v \sum_{p_1+} \cdots \sum_{+p_{2T}=s} \frac{\lambda_{v_1}^{p_1} \cdots \lambda_{v_{2T}}^{p_{2T}}}{p_1! \cdots p_{2T}!} E\left[(\nu_{v_1}^{2p_1} \cdots \nu_{v_{2T}}^{2p_{2T}})\boldsymbol{\nu}_v\boldsymbol{\nu}_v'\right] \mathbf{P}_v' \mathbf{L}_\Gamma'.$$

Notice that the (j^{th}, j^{th}) entry of the matrix $E\left[(\nu_{v1}^{2p_1} \cdots \nu_{v2T}^{2p_{2T}})\boldsymbol{\nu}_v \boldsymbol{\nu}_v'\right]$ is given by $E[\nu_{v1}^{2p_1}|\zeta_{v_1}, 1] \cdots E[\nu_{vj}^{2p_j+2}|\zeta_{vj}, 1] \cdots E[\nu_{v2T}^{2p_{2T}}|\zeta_{v2T}, 1]$, while the (l^{th}, j^{th}) entry is $E[\nu_{v1}^{2p_1}|\zeta_{v_1}, 1] \cdots E[\nu_{vl}^{2p_l+1}|\zeta_{vl}, 1] \cdots E[\nu_{vj}^{2p_j+1}|\zeta_{vj}, 1] \cdots E[\nu_{v2T}^{2p_{2T}}|\zeta_{v2T}, 1]$.

3.3 Separable Covariance Structure

Many of the statistical problems linked to the modeling of spatial-temporal dependence structures, can be overcome by using separable processes. In shape analysis, a major advantage of using a separable structure is that the covariance matrix can be decomposed (by means of a Kronecker product) into purely landmark and temporal components. This, dramatically reduces the number of parameters in the covariance matrix, facilitates computational procedures and also allows for the specification of commonly used temporal processes.

Assume that the $2TK \times 2TK$ covariance matrix, $\boldsymbol{\Sigma}^\dagger$, can be expressed as $\boldsymbol{\Sigma}^\dagger = \boldsymbol{\Sigma}_T \otimes \boldsymbol{\Sigma}_S^\dagger$, where $\boldsymbol{\Sigma}_T$ is a $T \times T$ covariance matrix between temporal observations and $\boldsymbol{\Sigma}_S^\dagger$ is a $2K \times 2K$ covariance matrix between landmark coordinates in the configuration space.

The covariance matrix in the pre-form space, $\boldsymbol{\Sigma} = (\mathbf{I}_{2T} \otimes \mathbf{L})\boldsymbol{\Sigma}^\dagger(\mathbf{I}_{2T} \otimes \mathbf{L}')$, is given by $\boldsymbol{\Sigma} = (\mathbf{I}_{2T} \otimes \mathbf{L})(\boldsymbol{\Sigma}_T \otimes \boldsymbol{\Sigma}_S^\dagger)(\mathbf{I}_{2T} \otimes \mathbf{L}') = \boldsymbol{\Sigma}_T \otimes [(\mathbf{I}_2 \otimes \mathbf{L})\boldsymbol{\Sigma}_S^\dagger(\mathbf{I}_2 \otimes \mathbf{L}')] = \boldsymbol{\Sigma}_T \otimes \boldsymbol{\Sigma}_S$. Therefore, $vec(\mathbf{X}) \sim \mathcal{N}_{2T(K-1)}(vec(\boldsymbol{\mu}), \boldsymbol{\Sigma}_T \otimes \boldsymbol{\Sigma}_S)$ has pdf

$$
\begin{aligned}
&f(vec(\mathbf{X})|\boldsymbol{\mu}, \boldsymbol{\Sigma}_T \otimes \boldsymbol{\Sigma}_S) \\
&= M \cdot exp\left\{-\frac{1}{2}[vec(\mathbf{X}) - vec(\boldsymbol{\mu})]'(\boldsymbol{\Sigma}_T^{-1} \otimes \boldsymbol{\Sigma}_S^{-1})[vec(\mathbf{X}) - vec(\boldsymbol{\mu})]\right\}
\end{aligned}
\tag{3.15}
$$

where $M = \left[(2\pi)^{(K-1)T}|\boldsymbol{\Sigma}_T|^{\frac{2(K-1)}{2}}|\boldsymbol{\Sigma}_S|^{\frac{T}{2}}\right]^{-1}$.

The $2(K-1) \times T$ matrix $\tilde{\mathbf{X}} = [vec(\mathbf{X}_1)\ vec(\mathbf{X}_2)\ \ldots\ vec(\mathbf{X}_T)]$ thus follows a matrix normal distribution, $\tilde{\mathbf{X}} \sim \mathcal{N}_{2(K-1),T}(\boldsymbol{\mu}, \boldsymbol{\Sigma}_S, \boldsymbol{\Sigma}_T)$, where $\boldsymbol{\mu} = [vec(\boldsymbol{\mu}_1)\ vec(\boldsymbol{\mu}_2)\ \ldots\ vec(\boldsymbol{\mu}_T)]$, $\boldsymbol{\Sigma}_S$ is the covariance among the rows of $\mathbf{X}$ (landmark coordinates in the pre-form space) and $\boldsymbol{\Sigma}_T$ is the covariance among the columns (temporal instants).

Given the induced shape distribution

$$
f(\mathbf{u}|\boldsymbol{\mu}, \boldsymbol{\Sigma}_T \otimes \boldsymbol{\Sigma}_S) = \frac{exp(-g/2)|\boldsymbol{\Gamma}|^{1/2}}{(2\pi)^{(K-2)T}|\boldsymbol{\Sigma}_T|^{\frac{2(K-1)}{2}}|\boldsymbol{\Sigma}_S|^{\frac{T}{2}}} \int \prod_{t=1}^{T} \|\mathbf{h}_t\|^{2(K-2)} f_{\mathcal{N}}(\mathbf{h}|\boldsymbol{\eta}, \boldsymbol{\Gamma})d\mathbf{h}
\tag{3.16}
$$

the covariance matrix $\boldsymbol{\Gamma}$, under the separability assumption, can be expressed in terms of the temporal and the spatial correlation structures as $\boldsymbol{\Gamma} =$

$\left(\mathbf{W}'(\boldsymbol{\Sigma}_T^{-1} \otimes \boldsymbol{\Sigma}_S^{-1})\mathbf{W}\right)^{-1}$ or equivalently as $\boldsymbol{\Gamma} = \left(\boldsymbol{\Sigma}_T^{-1} * \boldsymbol{\Gamma}_S^{-1}\right)^{-1}$ where $*$ represent the Katri-Rao product (Khatri and Rao 1968) between the inverse of the temporal covariance matrix and the inverse of the $2T \times 2T$ matrix $\boldsymbol{\Gamma}_S = \left(\mathbf{W}'(\mathbf{U}_T \otimes \boldsymbol{\Sigma}_S^{-1})\mathbf{W}\right)^{-1}$, with $\mathbf{U}_T = \mathbf{1}_t \otimes \mathbf{1}_t'$.

3.3.1 *EM for the Offset Shape Distribution of a Matrix-Variate Normal Distribution*

Considering the complete data $\mathscr{X} = \{\mathbf{X}^{(n)}\}_{n=1:N}$, in the preform space, and the rearrangement $\tilde{\mathbf{X}}^{(n)} = \left(vec(\mathbf{X}_1^{(n)}) \dots vec(\mathbf{X}_T^{(n)})\right)$, where $\tilde{\mathbf{X}}^{(n)} \sim \mathscr{N}_{2(K-1),T}(\boldsymbol{\mu}, \boldsymbol{\Sigma}_S, \boldsymbol{\Sigma}_T)$, $\forall n$, the ML estimators for $\boldsymbol{\Sigma}_S$ and $\boldsymbol{\Sigma}_T$, such that $\boldsymbol{\Sigma} = \boldsymbol{\Sigma}_T \otimes \boldsymbol{\Sigma}_S$, are given by (Dutilleul 1999)

$$\begin{aligned}\hat{\boldsymbol{\Sigma}}_S &= \frac{1}{NT}\sum_{n=1}^{N}(\tilde{\mathbf{X}}^{(n)} - \hat{\boldsymbol{\mu}})\hat{\boldsymbol{\Sigma}}_T^{-1}(\tilde{\mathbf{X}}^{(n)} - \hat{\boldsymbol{\mu}})' \\ &= \frac{1}{NT}\sum_{n=1}^{N}\tilde{\mathbf{X}}^{(n)}\hat{\boldsymbol{\Sigma}}_T^{-1}\tilde{\mathbf{X}}^{(n)\prime} - \frac{1}{T}\hat{\boldsymbol{\mu}}\hat{\boldsymbol{\Sigma}}_T^{-1}\hat{\boldsymbol{\mu}}' \qquad (3.17)\end{aligned}$$

$$\begin{aligned}\hat{\boldsymbol{\Sigma}}_T &= \frac{1}{N(2K-2)}\sum_{n=1}^{N}(\tilde{\mathbf{X}}^{(n)} - \hat{\boldsymbol{\mu}})'\hat{\boldsymbol{\Sigma}}_S^{-1}(\tilde{\mathbf{X}}^{(n)} - \hat{\boldsymbol{\mu}}) \\ &= \frac{1}{N(2K-2)}\sum_{n=1}^{N}\tilde{\mathbf{X}}^{(n)\prime}\hat{\boldsymbol{\Sigma}}_S^{-1}\tilde{\mathbf{X}}^{(n)} - \frac{1}{2K-2}\hat{\boldsymbol{\mu}}'\boldsymbol{\Sigma}_S^{-1}\hat{\boldsymbol{\mu}} \qquad (3.18)\end{aligned}$$

where $\hat{\boldsymbol{\mu}} = \frac{1}{N}\sum_{n=1}^{N}\tilde{\mathbf{X}}^{(n)}$.

In a ML framework (Mardia and Goodall 1993; Dutilleul 1999; Brown et al. 2001), these matrices are estimated iteratively as in a *flip-flop* algorithm (Lu and Zimmerman 2005). As known, the solutions for $\boldsymbol{\Sigma}_S$ and $\boldsymbol{\Sigma}_T$ are uniquely defined up to a scalar factor, but the estimate of the covariance matrix of the vectorial form is uniquely defined.

Given the Cholesky decompositions $\hat{\boldsymbol{\Sigma}}_T^{-1} = \mathbf{L}_T\mathbf{L}_T'$ and $\hat{\boldsymbol{\Sigma}}_S^{-1} = \mathbf{L}_S\mathbf{L}_S'$, Eqs. (3.17) and (3.18) can be written as

$$\hat{\boldsymbol{\Sigma}}_S = \frac{1}{NT}\sum_{n=1}^{N}\sum_{t=1}^{T}\check{\mathbf{A}}_t(\mathbf{L}_T' \otimes \mathbf{I}_{2K-2})vec(\mathbf{X}^{(n)})vec(\mathbf{X}^{(n)})'(\mathbf{L}_T \otimes \mathbf{I}_{2K-2})\check{\mathbf{A}}_t' - \frac{1}{T}\hat{\boldsymbol{\mu}}\hat{\boldsymbol{\Sigma}}_T^{-1}\hat{\boldsymbol{\mu}}' \qquad (3.19)$$

and

$$\hat{\boldsymbol{\Sigma}}_T = \frac{1}{N(2K-2)} \sum_{n=1}^{N} \sum_{k=1}^{2K-2} \breve{\mathbf{A}}_k (\mathbf{I}_T \otimes \mathbf{L}'_S) vec(\mathbf{X}^{(n)}) vec(\mathbf{X}^{(n)})' (\mathbf{I}_T \otimes \mathbf{L}_S) \breve{\mathbf{A}}'_k$$
$$- \frac{1}{2K-2} \hat{\boldsymbol{\mu}}' \hat{\boldsymbol{\Sigma}}_S^{-1} \hat{\boldsymbol{\mu}} \tag{3.20}$$

where $\breve{\mathbf{A}}_t = \boldsymbol{e}'_t \otimes \mathbf{I}_{2K-2}$, with the T-dimensional vector $\boldsymbol{e}_t$ defined as $e_t(t_1) = 1$, if $t = t_1$, and zero otherwise. Analogously, $\breve{\mathbf{A}}_k = \mathbf{I}_T \otimes \mathbf{e}'_k$, where the $(2K-2)$-dimensional vector $\mathbf{e}_k$ is such that $e_k(k_1) = 1$ if $k = k_1$ and zero otherwise.

Hence, conditional on the shape data $\mathscr{U} = \{\mathbf{u}^{(n)}\}_{n=1:N}$, parameter estimation can be performed through the EM algorithm, and the maximum of the conditional log-likelihood (M-step) is achieved at

$$vec(\boldsymbol{\mu}^{(r+1)}) = \frac{1}{N} \sum_{n=1}^{N} \int vec(\mathbf{X}^{(n)}) dF\left(\mathbf{X}^{(n)} | \mathbf{u}^{(n)}, \boldsymbol{\mu}^{(r)}, \boldsymbol{\Sigma}^{(r)}\right)$$

$$\boldsymbol{\Sigma}_S^{(r+1)} = \frac{1}{NT} \sum_{n=1}^{N} \sum_{t=1}^{T} \breve{\mathbf{P}}_t \int vec(\mathbf{X}^{(n)}) vec(\mathbf{X}^{(n)})' dF\left(\mathbf{X}^{(n)} | \mathbf{u}^{(n)}, \boldsymbol{\mu}^{(r)}, \boldsymbol{\Sigma}^{(r)}\right) \breve{\mathbf{P}}'_t +$$
$$- \frac{1}{T} \boldsymbol{\mu}^{(r+1)} \left(\boldsymbol{\Sigma}_T^{(r)}\right)^{-1} \boldsymbol{\mu}^{(r+1)'}$$

and

$$\boldsymbol{\Sigma}_T^{(r+1)} = \frac{1}{N(2K-2)} \sum_{n=1}^{N} \sum_{k=1}^{2K-2} \breve{\mathbf{P}}_k \int vec(\mathbf{X}^{(n)}) vec(\mathbf{X}^{(n)})' dF\left(\mathbf{X}^{(n)} | \mathbf{u}^{(n)}, \boldsymbol{\mu}^{(r)}, \boldsymbol{\Sigma}^{(r)}\right) \breve{\mathbf{P}}'_k +$$
$$- \frac{1}{2K-2} \boldsymbol{\mu}^{(r+1)'} \left(\boldsymbol{\Sigma}_S^{(r)}\right)^{-1} \boldsymbol{\mu}^{(r+1)}$$

where $\breve{\mathbf{P}}_t = \breve{\mathbf{A}}_t (\mathbf{L}_T^{(r)'} \otimes \mathbf{I}_{2K-2})$ and $\breve{\mathbf{P}}_k = \breve{\mathbf{A}}_k (\mathbf{I}_T \otimes \mathbf{L}_S^{(r)'})$.

The expectations of the sufficient statistics, given the observed data at current parameter estimates, $\boldsymbol{\mu}^{(r)}$ and $\boldsymbol{\Sigma}^{(r)} = \boldsymbol{\Sigma}_T^{(r)} \otimes \boldsymbol{\Sigma}_S^{(r)}$, can be computed considering Eqs. (3.8) and (3.9).

With regard to the landmark dependence structure, as seen in the previous chapter, we could assume a complex structure for $\boldsymbol{\Sigma}_S$. For the temporal dependence, the covariance matrix $\boldsymbol{\Sigma}_T$ could also be parameterized and a useful choice would be, for example, to assume the existence of a temporal autoregressive structure. However, even if we consider landmark isotropy, i.e. $\boldsymbol{\Sigma}_S = \mathbf{I}_{2(K-1)}$, $\boldsymbol{\Gamma} = (\mathbf{W}'(\boldsymbol{\Sigma}_T^{-1} \otimes \boldsymbol{\Sigma}_S^{-1})\mathbf{W}')^{-1}$ does not come in the form of a diagonal matrix, and the joint distribution of $\mathbf{h}$, $f_{\mathcal{N}_{2T}}(\mathbf{h}|\boldsymbol{\eta}, \boldsymbol{\Gamma})$, cannot be factorized in the product of its univariate marginal distributions. Hence, the estimation procedure does not simplify and remains computationally demanding. At this stage, simplifications of the procedure can only be achieved by assuming $\boldsymbol{\Sigma}_T$ as a multiple of the identity matrix and by modeling the dynamics of the process through a parameterized mean which captures the large-scale variability. In Sect. 3.4, we discuss the case in which the shape changes over time are modeled through a polynomial regression (Kume and Welling 2010).

3.3.2 Complex Landmark Covariance Structures

As shown in the previous chapter, second order circularity for the landmark covariance is achieved by assuming a complex structure for $\boldsymbol{\Sigma}_S$. Given the complex coordinates in the preform space and a separable covariance structure, let $\mathbf{Z}_t$, for each t, be a proper Gaussian complex random vector, such that the landmark variability is entirely characterized by the complex landmark covariance matrix $\boldsymbol{\Sigma}_{S_z} = 2(\mathbf{C}_{1,S} + i\mathbf{C}_{2,S})$ (see, Sect. 2.2). Under the separability hypothesis, it is straightforward to show that all the random configurations are jointly proper with complex covariance matrix $\boldsymbol{\Sigma}_z = \boldsymbol{\Sigma}_T \otimes \boldsymbol{\Sigma}_{S_z}$.

Given $vec(\mathbf{Z}) \sim \mathscr{CN}_{T(K-1)}(vec(\boldsymbol{\mu}_z), \boldsymbol{\Sigma}_T \otimes \boldsymbol{\Sigma}_{S_z})$, in order to derive the offset distribution of the shape coordinates, we consider the transformation $vec(\mathbf{Z}) = \boldsymbol{\Xi}\,\mathfrak{z}$, where $\boldsymbol{\Xi} = diag(\boldsymbol{\xi}_1 \dots \boldsymbol{\xi}_T)$ is a $T(K-1) \times T$ block-diagonal matrix of shape coordinates and $\mathfrak{z} = (z_{2,1} \dots z_{2,T})'$ is a T-dimensional vector of rotation and scale parameters. The joint distribution of $\boldsymbol{\xi}$ and $\mathfrak{z}$ is

$$
\begin{aligned}
f(\boldsymbol{\xi}, \mathfrak{z}|\boldsymbol{\mu}_z, \boldsymbol{\Sigma}_z) &= \frac{exp\left[-(\boldsymbol{\Xi}\,\mathfrak{z} - vec(\boldsymbol{\mu}_z))^* \boldsymbol{\Sigma}_z^{-1}(\boldsymbol{\Xi}\,\mathfrak{z} - vec(\boldsymbol{\mu}_z))\right] \prod_{t=1}^{T} \|z_{2,t}\|^{2(K-2)}}{\pi^{T(K-1)}|\boldsymbol{\Sigma}_z|} \\
&= \frac{exp(-g_z)}{\pi^{T(K-1)}|\boldsymbol{\Sigma}_z|} \prod_{t=1}^{T} \|z_{2,t}\|^{2(K-2)} exp\left[-(\mathfrak{z} - \mathfrak{n})^* \boldsymbol{\Gamma}_z^{-1}(\mathfrak{z} - \mathfrak{n})\right]
\end{aligned}
\tag{3.21}
$$

where $\boldsymbol{\Gamma}_z = (\boldsymbol{\Xi}^* \boldsymbol{\Sigma}_z^{-1} \boldsymbol{\Xi})^{-1}$, $\mathfrak{n} = \boldsymbol{\Gamma}_z \boldsymbol{\Xi}^* \boldsymbol{\Sigma}_z^{-1} vec(\boldsymbol{\mu}_z)$ and $g_z = vec(\boldsymbol{\mu}_z)^* \boldsymbol{\Sigma}_z^{-1} vec(\boldsymbol{\mu}_z) - \mathfrak{n}^* \boldsymbol{\Gamma}_z^{-1} \mathfrak{n}$.

The covariance matrix of $\mathfrak{z}$ can also be written as $\boldsymbol{\Gamma}_z = \left(\boldsymbol{\Sigma}_T^{-1} \odot \boldsymbol{\Gamma}_{S_z}^{-1}\right)^{-1}$, where $\odot$ is the Hadamard product and $\boldsymbol{\Gamma}_{S_z}^{-1} = \boldsymbol{\Xi}^*(\mathbf{U}_T \otimes \boldsymbol{\Sigma}_{S_z}^{-1})\boldsymbol{\Xi}$.

Since any subvector of a proper random vector is also proper (Neeser and Massey 1993), the T-dimensional complex-valued random vector $\mathfrak{z}$ has a proper complex Gaussian distribution, $\mathfrak{z} \sim \mathscr{CN}_T(\mathfrak{n}, \boldsymbol{\Gamma}_z)$, where $\boldsymbol{\Gamma}_z = 2(\mathbf{G}_1 + i\mathbf{G}_2)$.

Therefore the joint distribution, in Eq. (3.21), can be written as

$$f(\boldsymbol{\xi}, \mathfrak{z}|\boldsymbol{\mu}_z, \boldsymbol{\Sigma}_z) = \frac{exp(-g_z)|\boldsymbol{\Gamma}_z|}{\pi^{T(K-2)}|\boldsymbol{\Sigma}_z|} \prod_{t=1}^{T} (\mathfrak{z}^* \mathscr{A}_t \mathfrak{z})^{(K-2)} f_{\mathscr{C}\mathscr{N}}(\mathfrak{z}|\mathfrak{n}, \boldsymbol{\Gamma}_z) \tag{3.22}$$

where the $(T \times T)$ matrix $\mathscr{A}_t = \mathbf{e}_t'\mathbf{e}_t = diag(\mathbf{e}_t)$ has just the $\mathscr{A}(t,t) = 1$ entry different from zero.

The distribution of the shape variables is obtained by integrating out the rotation and scale parameters

$$\begin{aligned} f(\boldsymbol{\xi}|\mathfrak{z}, \boldsymbol{\mu}_z, \boldsymbol{\Sigma}_z) &= \frac{exp(-g_z)|\boldsymbol{\Gamma}_z|}{\pi^{T(K-2)}|\boldsymbol{\Sigma}_z|} \int \prod_{t=1}^{T} (\mathfrak{z}^* \mathscr{A}_t \mathfrak{z})^{(K-2)} f_{\mathscr{C}\mathscr{N}}(\mathfrak{z}|\mathfrak{n}, \boldsymbol{\Gamma}_z) d\mathfrak{z} = \\ &= \frac{exp(-g_z)|\boldsymbol{\Gamma}_z|}{\pi^{T(K-2)}|\boldsymbol{\Sigma}_z|} E\left[(\mathfrak{z}^H \mathscr{A}_t \mathfrak{z})^{(K-2)}\right] \end{aligned} \tag{3.23}$$

3.3.3 A Simulation Study

In this section, we describe results from some simulations we have conducted to investigate how the estimation procedure performs under the separable covariance structure. The studies were designed to give an indication of the performance of the EM algorithm, especially for different values of T and N. Specifically, the simulations were carried out using $K = 3$ (i.e. a triangle), $T = 4, 5, 6$, $N = 10, 20, 30$ and $\boldsymbol{\Sigma}_T$ having elements from the covariance structure of a first order autoregressive process with autoregressive parameter equal to 0.5. For simplicity, we have assumed here an isotropic structure for $\boldsymbol{\Sigma}_S^{\dagger}$, with $\sigma^2 = 1$.

Although the small values for K, N and T, estimation results suggest that the EM estimator performs reasonably well in terms of bias, standard errors and mean squared errors. Table 3.1 gives the results of estimation from 50 simulations. Note that the estimated mean shape of the triangle is given using Bookstein's coordinates and is thus represented by the pair of coordinates of the third landmark (i.e. $\mu_{3,1}$ and $\mu_{3,2}$). Also, notice that conditional on the values of $\boldsymbol{\mu}$, the autoregressive parameter is estimated numerically by minimizing the negative likelihood.

Estimation of $\boldsymbol{\Sigma}_T$, using Eq. (3.20), seems to be more difficult since, for the chosen parametrization, a bias can still be observed at the extreme values chosen for T and N. It also appears that the EM is not sensitive to the starting values, though some convergence problems were encountered in a few cases.[1]

The studies are not intended to be exhaustive and the choice of K, T and N was limited by the computational burden of the estimation procedure. To provide an idea

[1]Convergence problems appear to be more common by using a complex covariance structure for $\boldsymbol{\Sigma}_S^{\dagger}$

Table 3.1 The mean (and standard errors) of the parameter estimates from 50 simulations of a triangle with first-order autoregressive temporal structure

Parameters	$\hat{\mu}_{3,1}$	$\hat{\mu}_{3,2}$	$\hat{\rho}$
$T = 4, N = 10$	−0.177 (0.667)	−1.318 (0.564)	0.509 (0.146)
$T = 4, N = 20$	−0.217 (0.373)	−1.308 (0.350)	0.542 (0.127)
$T = 4, N = 30$	−0.212 (0.296)	−1.428 (0.321)	0.52 (0.111)
$T = 5, N = 10$	−0.162 (0.544)	−1.414 (0.583)	0.468 (0.148)
$T = 5, N = 20$	−0.211 (0.485)	−1.519 (0.475)	0.499 (0.120)
$T = 5, N = 30$	−0.212 (0.321)	−1.388 (0.329)	0.521 (0.080)
$T = 6, N = 10$	−0.157 (0.486)	−1.236 (0.434)	0.448 (0.162)
$T = 6, N = 20$	−0.162 (0.412)	−1.456 (0.399)	0.514 (0.097)
$T = 6, N = 30$	−0.189 (0.288)	−1.396 (0.327)	0.511 (0.060)

The true parameters are $\mu_{3,1} = -0.200$, $\mu_{3,2} = -1.400$ and $\rho = 0.5$

Table 3.2 CPU time (in seconds) required to compute the expectations in the E-step

	$K = 3$	$K = 4$
$T = 3$	0.26	6.8
$T = 4$	3.56	389.0
$T = 5$	54.0	21780.0
$T = 6$	825.0	–

Results refer to a single iteration of the EM algorithm, assuming $N = 1$

of the computational difficulties of working with Laguerre polynomials, Table 3.2 shows the CPU time[2] (in seconds) required to compute the expectations in Eqs. (3.8) and (3.9). Notice that results are for a single iteration of the EM algorithm, assuming $N = 1$, $K = 3, 4$ and T varying from 1 to 6. Table 3.2 suggests that is difficult to work with Laguerre polynomials with $K = 3$ and $T > 6$ and that, despite an explicit expression for $Q_s(\mathbf{B}_v)$ is available, this procedure is impractical for $K > 3$. One possibility to overcome these difficulties is to rely on Monte Carlo Integration, which is a simple and powerful technique for approximating complicated integrals such as those of Eqs. (3.8) and (3.9). In this case we only need to generate a set of random samples for the $\mathbf{h}'$s from the normal distribution with mean $\boldsymbol{\eta}$ and covariance matrix $\boldsymbol{\Gamma}$, i.e. $\mathbf{h} \sim \mathcal{N}_{2T}(\boldsymbol{\eta}, \boldsymbol{\Gamma})$. A simulation suggests that the expectations from Laguerre polynomials are reasonably approached by taking at least 5000 samples. Though this procedure can still be a time-consuming process, the use of Monte Carlo Integration would allow to work with configurations having $K > 3$.

[2]Results are obtained in Matlab with an Intel(R) Core(TM) i7-4558U CPU 2.80 GHz with 8 GB.

3.3.4 Temporal Independence

The existence of temporal correlation makes the procedure computationally demanding, limiting applications to a very small number of landmarks and temporal instants. Simplifications of the procedure can be achieved by assuming temporal independence.

Assuming $\boldsymbol{\Sigma}_T = \mathbf{I}_T$, the covariance matrix of the vector of scale and rotation parameters, $\boldsymbol{\Gamma}_S = \left(\mathbf{W}'(\mathbf{U}_T \otimes \boldsymbol{\Sigma}_S^{-1})\mathbf{W}\right)^{-1}$, has a block-diagonal structure, $\boldsymbol{\Gamma}^{-1} = \mathbf{I}_T * \boldsymbol{\Gamma}_S^{-1} = diag\left(\boldsymbol{\Gamma}_1^{-1}, \ldots, \boldsymbol{\Gamma}_T^{-1}\right)$, with $\boldsymbol{\Gamma}_t = \left(\mathbf{W}_t'\boldsymbol{\Sigma}_S^{-1}\mathbf{W}_t\right)^{-1}$.

From Eq. (3.3), the joint distribution of the shape variables and the rotation and scale parameters is now given by

$$\begin{aligned} f(\mathbf{h}, \mathbf{u}|\boldsymbol{\mu}, \boldsymbol{\Sigma}) &= \frac{exp(-g/2)}{(2\pi)^{(K-1)T}|\boldsymbol{\Sigma}_S|^{\frac{T}{2}}} \cdot \prod_{t=1}^{T} exp\left\{-\frac{(\mathbf{h}_t - \boldsymbol{\eta}_t)'\boldsymbol{\Gamma}_t^{-1}(\mathbf{h}_t - \boldsymbol{\eta}_t)}{2}\right\} \|\mathbf{h}_t\|^{2(K-2)} \\ &= \frac{|\boldsymbol{\Gamma}|^{\frac{1}{2}} exp(-g/2)}{(2\pi)^{(K-2)T}|\boldsymbol{\Sigma}_S|^{\frac{T}{2}}} \cdot \prod_{t=1}^{T} \|\mathbf{h}_t\|^{2(K-2)} f_{\mathcal{N}}(\mathbf{h}_t|\boldsymbol{\eta}_t, \boldsymbol{\Gamma}_t) \end{aligned} \quad (3.24)$$

where $|\boldsymbol{\Gamma}| = \prod_{t=1}^{T} |\boldsymbol{\Gamma}_t|$, and the marginal (off-set normal shape) pdf is

$$\begin{aligned} f(\mathbf{u}|\boldsymbol{\mu}, \boldsymbol{\Sigma}) &= \frac{|\boldsymbol{\Gamma}|^{\frac{1}{2}} exp(-g/2)}{(2\pi)^{(K-2)T}|\boldsymbol{\Sigma}_S|^{\frac{T}{2}}} \cdot \prod_{t=1}^{T} \int \|\mathbf{h}_t\|^{2(K-2)} f_{\mathcal{N}}(\mathbf{h}_t|\boldsymbol{\eta}_t, \boldsymbol{\Gamma}_t) d\mathbf{h}_t = \\ &= \frac{|\boldsymbol{\Gamma}|^{\frac{1}{2}} exp(-g/2)}{(2\pi)^{(K-2)T}|\boldsymbol{\Sigma}_S|^{\frac{T}{2}}} \prod_{t=1}^{T} E\left[(\mathbf{h}_t'\mathbf{h}_t)^{K-2}\right]. \end{aligned} \quad (3.25)$$

For each time t, the expected value $E\left[(\mathbf{h}_t'\mathbf{h}_t)^{K-2}\right]$ can be computed as seen in Sect. 2.4 and the EM update rules for the estimation of $\boldsymbol{\mu}$ and $\boldsymbol{\Sigma}_S$ are the same as those given in Sect. 2.5.

If in addition we assume a complex landmark covariance structure, we have $\boldsymbol{\Gamma}_z = diag(\gamma_{z,1}, \ldots, \gamma_{z,1})$ with $\gamma_{z,t} = \left(\boldsymbol{\xi}_t^* \boldsymbol{\Sigma}_{S_z}^{-1} \boldsymbol{\xi}_t\right)^{-1}$, and from equation (2.13), it follows that the joint distribution can be written as

$$\begin{aligned} f\left(\boldsymbol{\xi}, \mathfrak{z}|\boldsymbol{\mu}_z, \boldsymbol{\Sigma}_{S_z}\right) &= \prod_{t=1}^{T} f\left(\boldsymbol{\xi}_t, z_{2,t}|\boldsymbol{\mu}_{z,t}, \boldsymbol{\Sigma}_{S_z}\right) = \\ &= \frac{\prod_{t=1}^{T} \gamma_{z,t} \exp\{-g_{z,t}\}}{\pi^{T(K-2)} \left|\boldsymbol{\Sigma}_{S_z}\right|^{T}} \prod_{t=1}^{T} \|z_{2,t}\|^{2(K-2)} f_{\mathcal{CN}}(z_{2,t}|\eta_t, \gamma_{z,t}) \end{aligned} \quad (3.26)$$

where $\eta_t = \gamma_{z,t}\boldsymbol{\xi}_t^* \boldsymbol{\Sigma}_{S_z}^{-1}\boldsymbol{\mu}_{z,t}$ and $g_{z,t} = \boldsymbol{\mu}_{z,t}^* \boldsymbol{\Sigma}_{S_z}^{-1}\boldsymbol{\mu}_{z,t} - \bar{\eta}_t\gamma_{z,t}^{-1}\eta_t$. The offset-normal distribution is thus obtained as

$$f\left(\boldsymbol{\xi}|\boldsymbol{\mu}_z, \boldsymbol{\Sigma}_{S_z}\right) = \frac{\prod_{t=1}^{T}\gamma_{z,t}\exp\{-g_{z,t}\}}{\pi^{T(K-2)}\left|\boldsymbol{\Sigma}_{S_z}\right|^T}\prod_{t=1}^{T}\int \|z_{2,t}\|^{2(K-2)} f_{\mathscr{C}\mathscr{N}}(z_{2,t}|\eta_t, \gamma_{z,t}) =$$

$$= \frac{\prod_{t=1}^{T}\gamma_{z,t}\exp\{-g_{z,t}\}}{\pi^{T(K-2)}\left|\boldsymbol{\Sigma}_{S_z}\right|^T}\prod_{t=1}^{T}(K-2)!\gamma_{z,t}^{K-2}L_{K-2}\left(-\frac{\|\eta_t\|^2}{\gamma_{z,t}}\right) \tag{3.27}$$

where $L_{K-2}(\cdot)$ is the Laguerre polynomial of order $K-2$.

3.4 Offset Normal Distribution and Shape Polynomial Regression for Complex Covariance Structure

In this section we discuss a particular model to study the shape change in time. Specifically, we discuss a case in which the temporal dynamics of the shape variables are only modeled through a polynomial regression which captures the large-scale temporal variability of the process. It is thus assumed that $\boldsymbol{\Sigma}_T = \mathbf{I}_T$ and $\boldsymbol{\Sigma}_S$ is complex.

Supposing $vec(\mathbf{X}^\dagger) \sim N_{2KT}\left(vec(\boldsymbol{\mu}^\dagger), \boldsymbol{\Sigma}^\dagger\right)$, the mean of the process is parameterized by a polynomial function of order P, i.e. $\boldsymbol{\mu}_t^\dagger = E[\mathbf{X}_t^\dagger] = \sum_{p=0}^{P}\mathbf{B}_p^\dagger t^p$, with $\mathbf{B}_p^\dagger = \left(\boldsymbol{\beta}_p^{(x)^\dagger}\ \boldsymbol{\beta}_p^{(y)^\dagger}\right)$, and $\boldsymbol{\beta}_p^{(x)^\dagger}$ and $\boldsymbol{\beta}_p^{(y)^\dagger}$ K-dimensional vectors of regression coefficients.

Therefore, we can write $vec(\mathbf{X}^\dagger) \sim \mathscr{N}_{2KT}(\mathbf{D}^\dagger\boldsymbol{\beta}^\dagger, \boldsymbol{\Sigma}^\dagger)$, where $\boldsymbol{\beta}^\dagger = vec\left(\mathbf{B}_0^\dagger \ldots \mathbf{B}_P^\dagger\right)$ is a $2K(P+1)$-dimensional vector of regression coefficients, and the $(2KT \times 2K(P+1))$ design matrix can be constructed as $\mathbf{D}^\dagger = (\mathbf{T}\otimes\mathbf{I}_{2K})$, with

$$\mathbf{T} = \begin{pmatrix} 1 & t_1 & \ldots & t_1^P \\ 1 & t_2 & \ldots & t_2^P \\ \vdots & \vdots & \ldots & \vdots \\ 1 & t_T & \ldots & t_T^P \end{pmatrix}.$$

Notice that we require here that $T > P+1$. The coordinates in the preform space can be obtained as $vec(\mathbf{X}) = (\mathbf{I}_{2T}\otimes\mathbf{L})vec(\mathbf{X}^\dagger)$, and therefore $vec(\mathbf{X}) \sim \mathscr{N}_{2(K-1)T}(\mathbf{D}\boldsymbol{\beta}, \boldsymbol{\Sigma})$, with $\boldsymbol{\beta} = (\mathbf{I}_{2(P+1)}\otimes\mathbf{L})\boldsymbol{\beta}^\dagger$, $\mathbf{D} = \mathbf{T}\otimes\mathbf{I}_{2(K-1)}$, and $\boldsymbol{\Sigma} = (\mathbf{I}_{2T}\otimes\mathbf{L})\boldsymbol{\Sigma}^\dagger(\mathbf{I}_{2T}\otimes\mathbf{L}')$.

Given a random sample of N sequence of configurations, $\mathscr{X} = \{\mathbf{X}^{(n)}\}$, since the ML estimator of the regression parameters for the complete data is given by

$$\hat{\boldsymbol{\beta}} = \frac{1}{N}\sum_{n=1}^{N} \left(\mathbf{D}'\mathbf{D}\right)^{-1} \mathbf{D}' vec(\mathbf{X}^{(n)})$$

the update rule in the maximization step is

$$\boldsymbol{\beta}^{(r+1)} = \frac{1}{N}\sum_{n=1}^{N} \left(\mathbf{D}'\mathbf{D}\right)^{-1} \mathbf{D}' \int vec(\mathbf{X}^{(n)}) dF\left(\mathbf{X}^{(n)}|\mathbf{u}^{(n)}, \mathbf{D}\boldsymbol{\beta}^{(r)}, \boldsymbol{\Sigma}^{(r)}\right). \tag{3.28}$$

If we assume that the covariance has a Kronecker product form with temporal independence and that the configurations follow a proper complex Gaussian distribution, we have $vec(\mathbf{Z}^{\dagger}) \sim \mathscr{CN}_{KT}(\mathbf{D}_z^{\dagger}\boldsymbol{\beta}_z^{\dagger}, \mathbf{I}_T \otimes \boldsymbol{\Sigma}_{S_z}^{\dagger})$, where the $K(P+1)$-dimensional vector of complex regression coefficients is given by $\boldsymbol{\beta}_z^{\dagger} = vec\left(\boldsymbol{\beta}_0^{(x)\dagger} + i\boldsymbol{\beta}_0^{(y)\dagger} \dots \boldsymbol{\beta}_P^{(x)\dagger} + i\boldsymbol{\beta}_P^{(y)\dagger}\right)$, and the $K \times K(P+1)$ design matrix is constructed as $\mathbf{D}_z^{\dagger} = (\mathbf{T} \otimes \mathbf{I}_K)$. Consequently the distribution in the preform space is given by $vec(\mathbf{Z}) \sim \mathscr{CN}_{(K-1)T}(\mathbf{D}_z\boldsymbol{\beta}_z, \mathbf{I}_T \otimes \boldsymbol{\Sigma}_{S_z})$, with $\boldsymbol{\beta}_z = (\mathbf{I}_{P+1} \otimes \mathbf{L})\boldsymbol{\beta}_z^{\dagger}$, $\mathbf{D}_z = \mathbf{T} \otimes \mathbf{I}_{K-1}$, and $\boldsymbol{\Sigma}_{S_z} = \mathbf{L}\boldsymbol{\Sigma}_{S_z}^{\dagger}\mathbf{L}'$.

Now, in the maximization step of the EM algorithm, the update rules for the regression coefficients and the landmark covariance matrix are

$$\boldsymbol{\beta}_z^{(r+1)} = \frac{1}{N}\sum_{n=1}^{N} \tilde{\mathbf{D}}_z^{(r)} \int vec(\mathbf{Z}^{(n)}) dF\left(\mathbf{Z}^{(n)}|\boldsymbol{\xi}^{(n)}, \boldsymbol{\beta}_z^{(r)}, \boldsymbol{\Sigma}_{S_z}^{(r)}\right) \tag{3.29}$$

and

$$\boldsymbol{\Sigma}_{S_z}^{(r+1)} = \frac{1}{T}\sum_{t=1}^{T}\left[\frac{1}{N}\sum_{n=1}^{N}\int \mathbf{z}_t^{(n)}\mathbf{z}_t^{(n)*}\, dF\left(\mathbf{z}_t^{(n)}|\boldsymbol{\xi}_t^{(n)}, \boldsymbol{\beta}_z^{(r)}, \boldsymbol{\Sigma}_{S_z}^{(r)}\right) - \boldsymbol{\mu}_{z,t}^{(r+1)}\boldsymbol{\mu}_{z,t}^{(r+1)*}\right] \tag{3.30}$$

where $\tilde{\mathbf{D}}_z^{(r)} = \left(\mathbf{D}_z'\mathbf{D}_z\right)^{-1}\mathbf{D}_z'$, $\boldsymbol{\mu}_{z,t}^{(r+1)} = \mathbf{B}_z^{(r+1)}(1\ t \dots t^P)'$ and $\mathbf{B}_z = \left(\boldsymbol{\beta}_0^{(x)} + i\boldsymbol{\beta}_0^{(y)} \dots \boldsymbol{\beta}_P^{(x)} + i\boldsymbol{\beta}_P^{(y)}\right)$.

In the E-step of the algorithm, since $vec(\mathbf{Z}) = \boldsymbol{\Xi}\,\mathfrak{z}$ and $\mathbf{z}_t = \boldsymbol{\xi}_t z_{2,t}$, the expectations of the sufficient statistics can be computed as

$$\int vec(\mathbf{Z}) dF\left(\mathbf{Z}|\boldsymbol{\xi}, \boldsymbol{\beta}^{(r)}, \boldsymbol{\Sigma}_z^{(r)}\right) = \boldsymbol{\Xi}\frac{\int \mathfrak{z} f(\boldsymbol{\xi}, \mathfrak{z}|\boldsymbol{\beta}^{(r)}, \boldsymbol{\Sigma}_z^{(r)}) d\mathfrak{z}}{f(\boldsymbol{\xi}|\boldsymbol{\beta}^{(r)}, \boldsymbol{\Sigma}_z^{(r)})} \tag{3.31}$$

$$\int \mathbf{z}_t\mathbf{z}_t^* dF\left(\mathbf{z}_t|\boldsymbol{\xi}_t, \boldsymbol{\beta}^{(r)}, \boldsymbol{\Sigma}_{S_z}^{(r)}\right) = \boldsymbol{\xi}_t\frac{\int z_{2,t}\bar{z}_{2,t} f(\boldsymbol{\xi}_t, z_{2,t}|\boldsymbol{\beta}^{(r)}, \boldsymbol{\Sigma}_{S_z}^{(r)}) dz_{2,t}}{f(\boldsymbol{\xi}_t|\boldsymbol{\beta}^{(r)}, \boldsymbol{\Sigma}_{S_z}^{(r)})}\boldsymbol{\xi}_t^*$$

$$= \boldsymbol{\xi}_t \gamma_{z,t} (K-1) \left(\frac{\mathscr{L}_{K-1}\left(-\|\eta_t\|^2/\gamma_{z,t}\right)}{\mathscr{L}_{K-2}\left(-\|\eta_t\|^2/\gamma_{z,t}\right)} \right) \boldsymbol{\xi}_t^* \tag{3.32}$$

Since, as shown in Eqs. (3.26) and (3.27), $f(\boldsymbol{\xi}, \mathfrak{z}|\boldsymbol{\mu}_z, \boldsymbol{\Sigma}_{S_z}) = \prod_{t=1}^T f(\boldsymbol{\xi}_t, z_{2,t}|\boldsymbol{\mu}_{z,t}, \boldsymbol{\Sigma}_{S_z})$, and $f(\boldsymbol{\xi}|\boldsymbol{\mu}_z, \boldsymbol{\Sigma}_{S_z}) = \prod_{t=1}^T f(\boldsymbol{\xi}|\boldsymbol{\mu}_{z,t}, \boldsymbol{\Sigma}_{S_z})$, each element of the ratio in Eq. (3.31) can be computed as

$$\begin{aligned}
\frac{\int z_{2,l} f(\boldsymbol{\xi}, \mathfrak{z}|\boldsymbol{\beta}^{(r)}, \boldsymbol{\Sigma}_z^{(r)}) d\mathfrak{z}}{f(\boldsymbol{\xi}|\boldsymbol{\beta}^{(r)}, \boldsymbol{\Sigma}_z^{(r)})} &= \frac{\int z_{2,l} \prod_{t=1}^T f(\boldsymbol{\xi}_t, z_{2,t}|\boldsymbol{\mu}_{z,t}, \boldsymbol{\Sigma}_{S_z})\, d\mathfrak{z}}{\prod_{t=1}^T f(\boldsymbol{\xi}_t|\boldsymbol{\mu}_{z,t}, \boldsymbol{\Sigma}_{S_z})} \\
&= \frac{\int z_{2,l} f(\boldsymbol{\xi}_l, z_{2,l}|\boldsymbol{\mu}_{z,l}, \boldsymbol{\Sigma}_{S_z})\, dz_{2,l}}{f(\boldsymbol{\xi}_l|\boldsymbol{\mu}_{z,l}, \boldsymbol{\Sigma}_{S_z})} \\
&\quad \prod_{t \neq l} \frac{\int f(\boldsymbol{\xi}_t, z_{2,t}|\boldsymbol{\mu}_{z,t}, \boldsymbol{\Sigma}_{S_z})\, dz_{2,t}}{f(\boldsymbol{\xi}_t|\boldsymbol{\mu}_{z,t}, \boldsymbol{\Sigma}_{S_z})} \\
&= \frac{\int z_{2,l} f(\boldsymbol{\xi}_l, z_{2,l}|\boldsymbol{\mu}_{z,l}, \boldsymbol{\Sigma}_{S_z})\, dz_{2,l}}{f(\boldsymbol{\xi}_l|\boldsymbol{\mu}_{z,l}, \boldsymbol{\Sigma}_{S_z})} \\
&= \frac{\omega_t (K-1)}{\|\eta_t\|} \left(\frac{\mathscr{L}_{K-1}\left(-\|\eta_t\|^2/\gamma_{z,t}\right)}{\mathscr{L}_{K-2}\left(-\|\eta_t\|^2/\gamma_{z,t}\right)} - 1 \right)
\end{aligned}$$

where $\omega_t = e^{-i\theta}$, such that $\bar{\omega}_t \boldsymbol{\xi}_t^* \boldsymbol{\Sigma}_{S_z}^{-1} \boldsymbol{\mu}_{z,t}$ is a positive number. The solution of Eq. (3.31) thus follows from Sect. 2.5.1.

3.4.1 *Modeling the Dynamics of Facial Expressions by Shape Regression*

In this section we consider again the FG-NET data with the aim of using a polynomial shape regression to modeling the dynamics of the facial expressions. Differently from Sect. 2.6, we now work with all the 7 frames which have been chosen to summarize the dynamics of the expressions. Considering a complex covariance structure for $\boldsymbol{\Sigma}_S$, we fit both first and second order polynomial regression models. The standard AIC model selection criterion suggests that the model with $P = 2$ has to be preferred and the estimated mean paths for *happiness* and *surprise* expressions are shown in Fig. 3.1. The landmarks are represented by green dots while red dots represent the starting point of the estimated paths. The full estimated mean path is represented in blue.

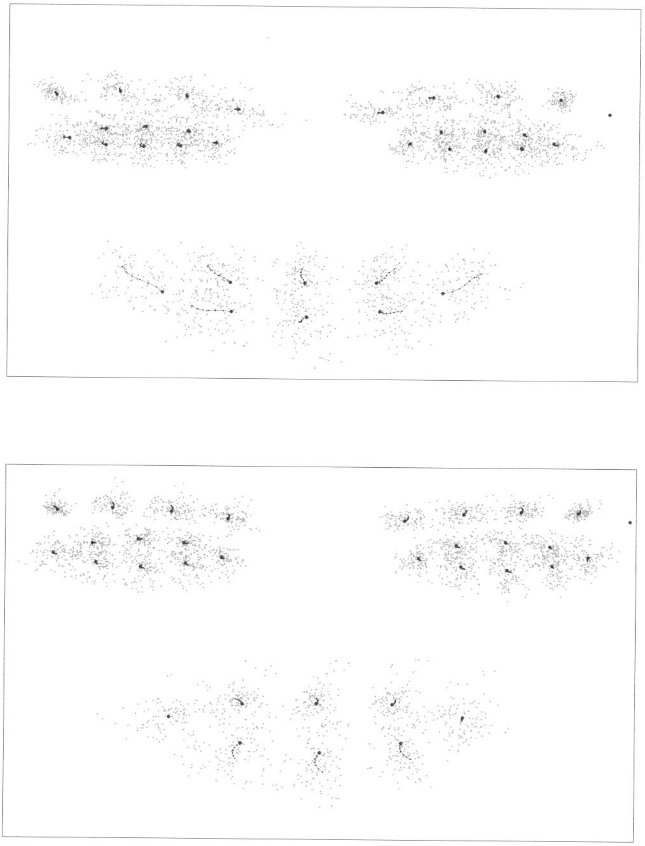

Fig. 3.1 Mean path estimates for *happiness* (*top*) and *surprise* (*bottom*) expressions

In general, it can be observed that *happiness* is characterized by a slight narrowing of the eyelids and a raising of the lip corners describing an upward curving of mouth and expansion on vertical and horizontal direction. On the other hand, *surprise* appears more with a vertical expansion of the mouth. As for *happiness*, the dynamics of the eyes, the eyelids and the eyebrows do not represent specific features of the expression.

3.5 Matching Symmetry

In various fields there is considerable interest in measuring bilateral symmetry of objects and in how to test the hypothesis of bilateral symmetry between left and right sides. In this section, for computational reasons, we only consider the problem of matching symmetry (Mardia et al. 2000) for which, given an object, we

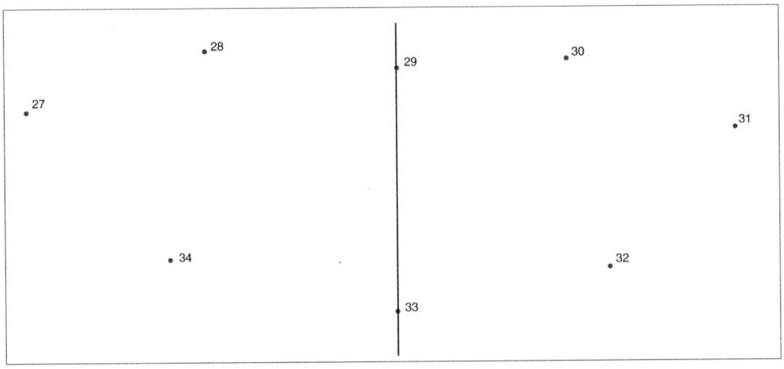

Fig. 3.2 Representation of the mouth region with the midline passing through the *solos* 29 and 33. In this region there are also the following three *paired* landmarks: (27, 31), (28, 30) and (34, 32)

have separate landmark configurations for the left and right sides that are mirror images of each other. For the FG-NET data, we constrain the analysis to the mouth region where, as shown in Fig. 3.2, we distinguish three *paired* landmarks and two *solos* which lie exactly on the midline of the mouth. In this case, for example, the *Happiness* expression is said symmetric if the pattern observed on the left side of the midline of the mouth is the same as the pattern observed on the right side. In order to verify the presence of symmetry in the mouth region, we thus compare the dynamics of the estimated mean paths of the triangle configurations represented by the landmarks on the left and right sides of the midline.

Let $\mathbf{X}^{\dagger(l)}$ and $\mathbf{X}^{\dagger(r)}$ be the left and right triangle configurations. We first reflect $\mathbf{X}^{\dagger(r)}$ using the orthogonal matrix, $\mathbf{H} = diag(-1, 1)$. Then, following Sect. 3.3.1, we estimate both the mean $\boldsymbol{\mu}$ and the covariance matrices, $\boldsymbol{\Sigma}_S$ and $\boldsymbol{\Sigma}_T$, for the left and right samples.

To test whether these mean paths differ from each other only by some rotation, as in Sect. 2.6, we may use the likelihood ratio test statistic with the hypothesis

$$\begin{cases} H_0 : \boldsymbol{\mu}_z^{(l)} = \boldsymbol{\mu}_z^{(r)} \\ H_1 : \boldsymbol{\mu}_z^{(l)} \neq \boldsymbol{\mu}_z^{(r)}, \quad \boldsymbol{\Sigma}_z^{(l)} = \boldsymbol{\Sigma}_z^{(r)}. \end{cases}$$

For the alternative hypothesis, the EM algorithm is applied to the left and right sides separately while keeping the entries of the covariance matrices, $\boldsymbol{\Sigma}_z^{(l)}$ and $\boldsymbol{\Sigma}_z^{(r)}$, equal to the estimated covariance matrix of the pooled sample. A schematic representation of the estimated mean paths for the pooled, left and right samples, is shown in Fig. 3.3. The estimated temporal covariance matrix suggests that, in average, the correlation between two-consecutive time points is around 0.45.

The resulting log-likelihood values for the left and right configurations are equal to 139.62 and 131.74, respectively. Instead, the log-likelihood for the pooled sample is 272.99. Since $-2 \log \Delta$ is distributed as χ_9^2 under the null hypothesis, the test suggests that modulo rotations, $\boldsymbol{\mu}_z^{(l)}$ and $\boldsymbol{\mu}_z^{(r)}$ are not different and hence, in the

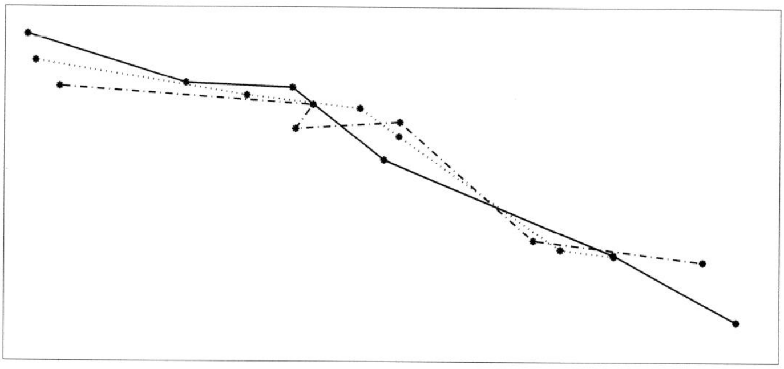

Fig. 3.3 Bookstein shape coordinates of the mean paths estimated for *left* (*continuous line*) and *right* (*dashed line*) triangle configurations. The *dot-dashed* path represents the dynamics of the pooled sample. The paths are obtained by estimating the parameters of a Matrix-Variate Normal distribution for the three paired landmarks within the mouth region

mouth region, the *Happiness* expression is symmetric in its dynamics. Of course the test assumes that the left- and right-side samples should be independent and this might not be true in practice.

Notice that here, for computational reasons, we have restricted the analysis to only 3 landmarks. However, the NonParametric Combination (NPC) test is able to work with the complete landmark configuration and can also consider the problem of bilateral symmetry in terms of object symmetry (Mardia et al. 2000). As it will be shown in Chap. 5, results from this analysis support those achieved here for the mouth region.

3.6 Mixture Models for Classification

In many applications of data modeling finite mixtures of distributions (see, for example, McLachlan and Peel 2000 and Everitt and Hand 1981) provide a sensible model for a data at hand. Because of their flexibility and usefulness, they have continued to receive increasing attention over the years.

In this section we discuss the use of a mixture of offset-normal shape distributions for classification problems. In practice, for *happiness* and *surprise* expressions, we only consider a two-component problem and, as in previous section, we concentrate the analysis on the mouth region which is the part showing the most significant temporal changes.

Assume that $\mathbf{X}^{\dagger^{(1)}}, \ldots, \mathbf{X}^{\dagger^{(N)}}$ is a sample from a mixture of G normal distributions with density $f_{\mathcal{N}}\left(\mathbf{X}^{\dagger}\right) = \sum_{g=1}^{G} \pi_g f_{\mathcal{N},g}\left(\mathbf{X}^{\dagger}|\boldsymbol{\mu}_g^{\dagger}, \boldsymbol{\Sigma}_g^{\dagger}\right)$, where $\pi_g \geq 0$ and $\sum_g^G \pi_g = 1$. Clearly the induced pdf in the preform space is $f_{\mathcal{N}}(\mathbf{X}) = \sum_{g=1}^{G} \pi_g f_{\mathcal{N},g}\left(\mathbf{X}|\boldsymbol{\mu}_g, \boldsymbol{\Sigma}_g\right)$ and the induced distribution of shape variables is given by a linear combination of offset-normal distributions, that is $f(\mathbf{u}) = \sum_{g=1}^{G} \pi_g f_g\left(\mathbf{u}|\boldsymbol{\mu}_g, \boldsymbol{\Sigma}_g\right)$.

The problem of estimating the parameters in a finite mixture has been studied extensively in literature. The book by Everitt and Hand (1981) provides an excellent overview of this topic and offers several methods for parameter estimation. In general, model parameter estimates can be obtained by maximizing the log-likelihood function for the entire set of N (independent) configurations, and the EM algorithm is a standard tool in this framework. In statistical shape analysis, considering a sample of configurations at a specific time, Kume and Welling (2010) provide the relevant update rules of the EM algorithm for the parameters π_g, $\boldsymbol{\mu}_g$ and $\boldsymbol{\Sigma}_g$. Here, these update rules are extended to a dynamic setting where, as in Sect. 3.4, we assume that the mean can be parameterized by a polynomial function. Furthermore we consider a separable covariance structure with independence in time and a complex covariance structure for the landmarks. The update rules are given below in complex notation.

Given the number G of mixture components, the estimated posterior probability that an observed sequence of shapes, $\boldsymbol{\xi}^{(n)}$, belongs to the g-th term is given by

$$P\left(g|\boldsymbol{\xi}^{(n)}\right) = \frac{\pi_g^{(r)} f_g\left(vec\left(\boldsymbol{\xi}^{(n)}\right)|\boldsymbol{\mu}_{z,g}^{(r)}, \boldsymbol{\Sigma}_{z,g}^{(r)}\right)}{\sum_{j=1}^{G} \pi_j^{(r)} f_j\left(vec\left(\boldsymbol{\xi}^{(n)}\right)|\boldsymbol{\mu}_{z,j}^{(r)}, \boldsymbol{\Sigma}_{z,j}^{(r)}\right)}, \quad g = 1, \ldots, G;\ n = 1, \ldots, N$$

where, as in Sect. 3.4, $\boldsymbol{\mu}_{z,g}^{(r)} = \mathbf{D}_z \boldsymbol{\beta}_{z,g}^{(r)}$ and $\boldsymbol{\Sigma}_{z,g}^{(r)} = \mathbf{I}_T \otimes \boldsymbol{\Sigma}_{S_z,g}^{(r)}$.

We can use the estimated posterior probabilities to obtain a weighted update of the parameters for each component. This gives the iterative EM update equations for the mixing coefficients, the parameterized means and the covariance matrices:

$$\pi_g^{(r+1)} = \frac{1}{N} \sum_{n=1}^{N} P\left(g|\boldsymbol{\xi}^{(n)}\right)$$

$$\boldsymbol{\beta}_{z,g}^{(r+1)} = \frac{1}{\sum_{n=1}^{N} P\left(g|\boldsymbol{\xi}^{(n)}\right)} \sum_{n=1}^{N} P\left(g|\boldsymbol{\xi}^{(n)}\right) \tilde{\mathbf{D}}_{z,g}^{(r)}$$

$$\int vec\left(\mathbf{Z}^{(n)}\right) dF\left(\mathbf{Z}^{(n)}|\boldsymbol{\xi}^{(n)}, \boldsymbol{\beta}_{z,g}^{(r)}, \boldsymbol{\Sigma}_{z,g}^{(r)}\right)$$

$$\boldsymbol{\Sigma}_{S_z,g}^{(r+1)} = \frac{1}{T} \sum_{t=1}^{T} \left[\frac{1}{\sum_{n=1}^{N} P\left(g|\boldsymbol{\xi}^{(n)}\right)} \sum_{n=1}^{N} \int \mathbf{z}_t^{(n)} \mathbf{z}_t^{(n)*} dF \right.$$

$$\left. \left(\mathbf{z}_t^{(n)}|\boldsymbol{\xi}_t^{(n)}, \boldsymbol{\beta}_{z,g}^{(r)}, \boldsymbol{\Sigma}_{S_z,g}^{(r)}\right) - \boldsymbol{\mu}_{z,t,g}^{(r+1)} \boldsymbol{\mu}_{z,t,g}^{(r+1)*} \right]$$

where $\tilde{\mathbf{D}}_{z,g}^{(r)} = \left(\mathbf{D}_{z,g}' \mathbf{D}_{z,g}\right)^{-1} \mathbf{D}_{z,g}'$, and $\boldsymbol{\mu}_{z,t,g}^{(r+1)} = \mathbf{B}_{z,g}^{(r+1)} (1\ t \ldots t^P)'$ with $\mathbf{B}_{z,g} = \left(\boldsymbol{\beta}_{z,0,g}^{(x)} + i\boldsymbol{\beta}_{z,0,g}^{(y)} \ldots \boldsymbol{\beta}_{z,P,g}^{(x)} + i\boldsymbol{\beta}_{z,P,g}^{(y)}\right)$.

Table 3.3 Confusion matrices for the *happiness* and *surprise* expressions with first and second order polynomials for the mean functions and complex covariance structure for the landmarks

		Happiness	Surprise
P=1	Happiness	12	4
	Surprise	0	16
P=2	Happiness	15	1
	Surprise	2	14

Results on hypothesis testing performed in Sect. 2.6 provided some evidence about significant differences between *happiness* and *surprise* expressions. Here, by using first and second order polynomials for the mean functions and a complex covariance structure for the landmarks, we use mixture models (assuming $G = 2$) for classifying the two facial expressions. Table 3.3, which represents the confusion matrices for the estimated models, shows that with a misclassification error of about 6.0%, obtained for the second order polynomial regression, the two expressions can be easily distinguished.

References

Brown P, Kenward M, Bassett E (2001) Bayesian discrimination with longitudinal data. Biostatistics 2:417–432

Dutilleul P (1999) The mle algorithm for the matrix normal distribution. J Stat Comput Simul 64:105–123

Everitt B, Hand D (1981) Finite mixture distributions. Chapman & Hall, London

Fishbaugh J, Prastawa M, Durrleman S, Piven J, Gerig G (2012) Analysis of longitudinal shape variability via subject specific growth modeling. In: Ayache N, Delingette H, Golland P, Mori K (eds) Medical image computing and computer-assisted intervention – MICCAI 2012. Lecture notes in computer science, vol 7510. Springer, Berlin, Heidelberg, pp 731–738

Fontanella L, Ippoliti L, Valentini P (2013) A functional spatio-temporal model for geometric shape analysis. In: Torelli N, Pesarin F, Bar-Hen A (eds) Advances in theoretical and applied statistics. Springer, Berlin, pp 75–86

Hinkle J, Muralidharan P, Fletcher PT, Joshi S (2012) International anthropometric study of facial morphology in various ethnic groups/races. In: Computer Vision - ECCV. Lecture Notes in Computer Science, vol 7574. pp 1–14

Kan R (2008) From moments of sum to moments of product. J Multivar Anal 99(3):542–554

Kent JT, Mardia KV, Morris RJ, Aykroyd RG (2001) Functional models of growth for landmark data. In: Proceedings in functional and spatial data analysis. University Press, Leeds, pp 109–115

Khatri C, Rao C (1968) Solutions to some functional equations and their applications to characterization of probability distributions. Sankhya 30:167–180

Kume A, Welling M (2010) Maximum likelihood estimation for the offset-normal shape distributions using em. J Comput Graph Stat 19:702–723

Kume A, Dryden I, Le H (2007) Shape space smoothing splines for planar landmark data. Biometrika 94:513–528

Lu N, Zimmerman D (2005) The likelihood ratio test for a separable covariance matrix. Stat Probab Lett 73:449–457

Magnus J (1986) The exact moments of a ratio of quadratic forms in normal variables. Ann Econ Stat 4:95–109

Mardia K, Goodall C (1993) Spatial-temporal analysis of multivariate environmental monitoring data. In: Patil G, Rao C (eds) Multivariate environmental statistics, vol 6. North-Holland, New York, pp 347–385

Mardia KV, Walder AN (1994) Shape analysis of paired landmark data. Biometrika 81:185–196

Mardia KV, Bookstein FL, Moreton IJ (2000) Statistical assessment of bilateral symmetry of shapes. Biometrika 87:285–300

Mathai A, Provost S (1992) Quadratic forms in random variables: theory and applications. Dekker, New York

McLachlan GJ, Peel D (2000) Finite mixture models. Wiley, New York

Neeser F, Massey J (1993) Proper complex random processes with applications to information theory. IEEE Trans Inf theory 39(4):1293–1302

Part II
Combination-Based Permutation Tests for Shape Analysis

Chapter 4
Parametric and Non-parametric Testing of Mean Shapes

Abstract This chapter deals with inferential aspects in shape analysis. At first we review inferential methods known in the shape analysis literature, highlighting some drawbacks of using Hotelling's T^2 test statistic. Then we present an extension of the NonParametric Combination (NPC) methodology to compare shape configurations of landmarks.

NPC tests represent an appealing alternative since they are distribution-free and allow for quite efficient solutions when the number of cases is lower than the number of variables (i.e., (semi)landmarks coordinates). This allows to obtain better representations of shapes even in presence of small sample size. NPC methodology enables to provide global as well as local evaluation of shapes: it is then possible to establish whether in general two shapes are different and which landmark/subgroup of landmarks mainly contributes to differentiate shapes under study. NPC tests enjoy the *finite-sample consistency* property hence, in this nonparametric framework, it is possible to obtain efficient solutions for multivariate small sample problems, like those encountered in the shape analysis field. We finally present a NPC solution for longitudinal data.

Keywords Multi-Aspect approach • NonParametric Combination methodology • Permutation tests combination-based for repeated measures design

4.1 Inferential Procedures for the Analysis of Shapes

The statistical community has shown an increased interest in shape analysis in the last decade and particular efforts have been addressed to the development of powerful statistical methods based on models measuring the shape variation of entire landmark configurations. Rohlf (2000) reviews the main tests used in the field of shape analysis and compares the statistical power of various tests that have been proposed to test for equality of shape in two populations. Even if his work is limited to the simplest case of homogeneous, independent, spherical variation at each landmark and the sampling experiments emphasize the case of triangular shapes, it allows the practitioners to choose the method that has the highest statistical power under a set of assumptions that are appropriate for the data. Through a

C. Brombin et al., *Parametric and Nonparametric Inference for Statistical Dynamic Shape Analysis with Applications*, SpringerBriefs in Statistics,
DOI 10.1007/978-3-319-26311-3_4

simulation study, he found that Goodall's F-test had the highest power followed by T^2-test using Kendall tangent space coordinates. Power for T^2-tests using Bookstein shape coordinates was good if the baseline was not the shortest side of the triangle. The Rao and Suryawanshi shape variables had much lower power when triangles were not close to being equilateral. With reference to interlandmark distance-based approaches, power surfaces for the `EDMA-I` T statistic revealed very low power for many shape comparisons including those between very different shapes. We remind the reader that EDMA stands for Euclidean Distance Matrix Analysis. On the other hand, power surface for the `EDMA-II` Z statistic depended strongly on the choice of baseline used for size scaling (Rohlf 2000).

All the above mentioned tests are based on quite stringent assumptions. In particular, the tests based on the T^2 statistic (e.g. T^2-tests using Bookstein, Kendall tangent space coordinates, Rao and Suryawanshi shape variables, like Rao-d (1996) and Rao-a (1998)) require independent samples, homogeneous covariance matrices and shape coordinates distributed according to the multivariate normal distribution. We remark that Hotelling's T^2 test statistic is derived under the assumption of population multivariate normality and it may not be very powerful unless there are a large number of observations available (Dryden and Mardia 1998). It is well known in the literature that Hotelling's T^2 test is formulated to detect any departures from the null hypothesis and therefore often lacks power to detect specific forms of departures that may arise in practice, i.e. the T^2 test fails to provide an easily implemented one-sided (directional) hypothesis test (Blair et al. 1994).

Goodall's F test requires a restrictive isotropic model and assumes that the distributions of the squared Procrustes distances are approximately Chi-squared distributed.

If we consider the methods based on interlandmark distances, `EDMA-I` T assumes independent samples and the equality of the covariance matrices in the two populations being compared (Lele and Cole 1996), while `EDMA-II` Z assumes only independent samples and normally distributed variation at each landmark.

In order to complete the review on main tests used in shape analysis, we recall the development and application of bootstrap methods in this field. In particular, we mention the pivotal bootstrap methods for multisample problems with directional data or shape data, proposed in the paper by Amaral et al. (2007). The basic assumption here is that the distribution of the sample mean shape (or direction or axis) is highly concentrated. This is substantially weaker assumption than is entailed in tangent space inference (Dryden and Mardia 1998) where observations are presumed highly concentrated.

Authors presented an extensive simulation study to investigate the performance of the proposed $\lambda_{\min}$ statistic (i.e., the smallest eigenvalue of a certain positive matrix), Goodall, Hotelling, and James statistics. Simulation results showed that bootstrap procedure performs better than parametric procedures in various situations and may be used to analyze landmark-based shapes in 3 or more dimensions (Dryden et al. 2008; Preston and Wood 2010, 2011).

As pointed out in Good (2000), the assumption of equal covariance matrices may be unreasonable especially under the alternative, the multinormal model in the tangent space may be doubted and sometimes there are few individuals and many landmarks, implying over-dimensioned spaces and loss of power for the Hotelling's T^2 test. Hence when sample sizes are too small, or the number of landmarks is too large, it is essentially inefficient to assume that observations are normally distributed. An alternative procedure is to consider a permutation version of the test (see Good 2000; Dryden and Mardia 1993; Bookstein 1997; Terriberry et al. 2005). Permutation methods are distribution-free, allow us for quite efficient solutions when the number of cases is less than the number of covariates and may be tailored for sensitivity to specific treatment alternatives providing one-sided as well as two-sided tests of hypotheses (Blair et al. 1994).

On the basis of these considerations, an extension of the NonParametric Combination (NPC) methodology has been proposed (Pesarin 2001; Pesarin and Salmaso 2010b; Brombin 2009; Brombin et al. 2008; Brombin and Salmaso 2009; Brombin et al. 2009a,b; Alfieri et al. 2012).

In particular in Brombin and Salmaso (2009), an exhaustive simulation study has been carried out to compare power behaviour of traditional tests proposed in literature with that of the NPC tests. Actually, traditional Hotelling's T^2 test may not be very powerful in presence of small samples (Dryden and Mardia, 1998, Blair et al., 1994). For these reasons, a nonparametric permutation counterpart has been proposed and it has been shown that the power of this test increases when increasing the number of the analyzed variables, even when the number of analyzed variables is larger than the permutation sample space.

On the basis of these results, throughout a simulation study, it has been illustrated that power of multivariate NPC tests increases when increasing the number of the processed variables provided that the noncentrality parameter increases, even when the number of covariates is larger than the permutation sample space. This behavior reflects the notion of *finite-sample consistency* for permutation tests combination-based. Specifically, for a given and fixed number of subjects, when the number of variables and the associated noncentrality parameter, induced by the test statistic, both diverge, then the power function of multivariate NPC tests based on associative statistics converges to one.

These results hold true even when considering functions of the noncentrality parameter or in presence of random effects.

4.2 NPC Approach in Shape Analysis

4.2.1 Brief Description of the Nonparametric Methodology

The NonParametric Combination (NPC) methodology (Pesarin 2001; Pesarin and Salmaso 2010b) is a conditional testing procedure that, under very mild and reasonable conditions, provided that exchangeability of data with respect to groups is satisfied in the null hypothesis, is found to be consistent and unbiased (Celant et al. 2000; Pesarin and Salmaso 2010a).

An extension of the NPC methodology to shape analysis was originally proposed by Brombin (2009) and Brombin and Salmaso (2013).

NPC tests are relatively efficient and do not require strong underlying assumptions as the traditional parametric competitors or standard distribution-free methods based on ranks, which are generally not conditional on sufficient statistics and almost never show better unconditional power behaviour.

Actually, permutation tests are essentially exact in a nonparametric conditional framework, where conditioning is on the pooled observed data set, which is generally a set of sufficient statistics in the null hypothesis.

Provided the permutation principle applies, one major feature of the nonparametric combination of dependent tests is that attention must be paid to a set of partial tests, each appropriate for the related sub-hypotheses. In general, the researcher is not explicitly required to specify the dependence structure of response variables since the underlying dependence structure is nonparametrically and implicitly captured by the combining procedure. Moreover, in this framework conditional inferential results may be extended to the unconditional ones (Pesarin 2002) and, due to their nonparametric nature, NPC tests may be computed even when the number of covariates exceeds the number of cases.

The NPC methodology consists of the following steps:

- a *breaking down* of the hypotheses into r, $r > 1$, sub-hypotheses, where for each sub-hypothesis, a suitable partial permutation test statistic is available;
- a *conditional simulation procedure* which, by conditioning with respect to the set of observed data, provides an estimate of the null multivariate permutation distribution of the whole set of test statistics;
- a *combination of the partial tests* into a second-order statistic whose null permutation distribution is estimated by using the same simulation results of the previous step (Celant et al. 2000).

In the context of shape analysis, the breaking down of the hypotheses enables to provide global as well as local evaluation of shapes: it is then possible to establish whether in general two shapes are different and which landmarks, or subgroups of landmarks (i.e. *domains*), mainly contribute to differentiate the shapes under study.

With K landmarks in m dimensions, by applying the NPC methodology, the hypothesis testing problem is broken down into two stages, considering both coordinates and landmark levels. Partial test statistics for one-sided hypotheses can be formulated and a global test then follows by combining at the first stage coordinates with respect to m and then with respect to K.

Hence, in two dimensions (i.e. $m = 2$), shape coordinates, give rise to the sub-hypotheses of the problem and thus provide the basis for a set of partial tests (namely, coordinate partial tests). By combining these partial tests, it is then possible to obtain a p-value for each landmark. Depending on the problem at hand, one could focus on the coordinate level or on the landmark level (after combining coordinates) and, finally, on the global test.

Using the NPC methodology a researcher is thus able to obtain not only a global p-value, as in traditional tests, but also a p-value for each landmark. Partial tests can provide marginal information for each specific landmark while jointly they can provide information on the global hypothesis. In this way, if we find a significant departure from the null hypothesis, one can investigate the nature of this departure in detail.

It should be noted that proceeding in this way, multiplicity problems may arise due to the large number of hypotheses to be tested on the same data. Hence intermediate partial p-values need to be adjusted for multiplicity.

Among all the "good" properties of NPC tests, we mention the finite-sample consistency (FSC) notion (Pesarin and Salmaso 2010a). For a given and fixed number of subjects, when the number of variables (e.g., landmark coordinates) and the associated noncentrality parameter, induced by the test statistic, both diverge, then the power function of multivariate NPC tests based on associative statistics converges to one.

Such findings look very useful to solve multivariate small sample problems, often occurring in shape analysis. Most of traditional inferential methods in shape analysis are parametric and they often require large sample size while, in practice, researchers may have to work with *fat data*, where there are more variables than observations.

Many complex multivariate problems, like those faced when dealing with data of two- or three-dimensional shapes/objects, are difficult to handle outside the conditional framework and in particular outside the nonparametric combination (NPC) of dependent permutation tests method. As pointed out in Pesarin and Salmaso (2010b), despite in the literature permutation tests are mostly derived by means of heuristic arguments (Edgington and Onghena 2007; Good 2005), their natural theoretical background must be referred to the principles of conditional inference (Birnbaum 1962; Edwards 1972). Since within this framework it can be proved that permutation tests are provided with suitable theoretical properties (Pesarin and Salmaso 2010b, 2012), whenever permutation tests are correctly applicable, their results may be extended, at least in a weak sense, to population inferences (Pesarin 2002).

It is worth noting that within a parametric framework the extension from samples to populations is possible only when the data set is randomly selected by well-designed sampling procedures on well-defined population distributions, provided that their nuisance parameters are completely removable (Pesarin 2002). When these conditions fail, especially if selection-bias procedures are used for data collection processes, in general most of the parametric inferential extensions are wrong or misleading. On the contrary, the permutation-based inferential conclusions may be always extended to the reference population even in case of selection-bias sampling (Pesarin 2002).

4.2.2 A Two Independent Sample Problem with Landmark Data

Let us assume that several landmark data $\{X^{\dagger}_{gij_s}\}$ are collected on different individuals grouped into sensible groupings, where $i = 1, \ldots, n_g$ labels different individuals, $g = 1, \ldots, G$ labels the groups in which subject i belongs to, $j = 1, \ldots, K$ labels different landmarks, $s = 1, \ldots, m$ labels different dimensions.

Without loss of generality, let us examine a two independent sample problem $(g = 1, 2, n_1, n_2)$ where objects are represented in the plane $(m = 2)$.

Let $\{\mathbf{V}_{g_i}\}$ denote the $K \times 2$ (centered not Helmertized) matrix of Procrustes tangent coordinates of the data $\{\mathbf{X}^{\dagger}_{g_i}\}$.

In practice, denoting by $(a_1^*, \ldots, a_N^*)$ a permutation of the labels $(1, \ldots, N)$, $\{\mathbf{V}_{g_i}\}^* = \{V^*_{gij_s} = V_{g_{a_i^*}j_s}, i = 1, \ldots, n_g, g = 1, 2, j = 1, \ldots, K, s = 1, 2\}$ is the related permutation of $\mathbf{V}_{g_i}$, so that $V^*_{1j_s} = \{V^*_{1ij_s} = V_{1_{a_i^*}j_s}, i = 1, \ldots, n_1, j = 1, \ldots, K, s = 1, 2\}$ and $V^*_{2j_s} = \{V^*_{2ij_s} = V_{2_{a_i^*}j_s}, i = n_1 + 1, \ldots, N, j = 1, \ldots, K, s = 1, 2\}$ are the two permuted samples, respectively.

For simplicity, we may assume that the landmark coordinates in tangent space behave according to the following model:

$$V_{j_s g_i} = \mu_{j_s} + \delta_{gj_s} + \sigma_{j_s} Z_{gij_s},$$

$i = 1, \ldots, n_g, g = 1, 2, j = 1, \ldots, K, s = 1, 2$, where

- μ_{j_s} represents a population constant for the j_sth variable (i.e. landmark coordinate);
- δ_{gj_s} represents a group effect at level g on the j_sth variable which, without loss of generality, is assumed to be $\delta_{1j_s} = 0$, $\delta_{2j_s} \leq$ (or $\geq$) 0, $\forall (j_s)$;
- σ_{j_s} represent population scale coefficients for variable j_s;
- Z_{gij_s} are random errors assumed to be exchangeable with respect to treatment levels, independent with respect to units, with zero mean, $E(\mathbf{Z}) = \mathbf{0}$, and finite second moment (Pesarin and Salmaso 2010b).

Hence landmark coordinates in the first group differ from those in the second group by a 'quantity' δ, where δ represents a q-dimensional vector of effects, with $q = K \times 2$. Again, $V^*_{gij_s}$, $i = 1, \ldots, n_g$, $g = 1, 2, j = 1, \ldots, K$, $s = 1, 2$, indicates a permutation of the original data.

Therefore the specific hypotheses may be expressed as

$$H_0 : \bigcap_{j_s=1}^{q} \{V_{1j_s} \overset{d}{=} V_{2j_s}\} \quad \text{vs.} \quad H_1 : \bigcup_{j_s}^{q} \{(V_{1j_s} + \delta) \overset{d}{>} V_{2j_s}\},$$

where $\overset{d}{>}$ stands for distribution (or stochastic) dominance.

With $T^o_{j_s}(0)$ and $T^*_{j_s}(0)$ we indicate respectively the observed and permutation values of T_{j_s} when $\delta = 0$, i.e. under H_0.

The assumptions regarding the set of partial tests $\mathbf{T} = \{T_{j_s}, j = 1, \ldots, K, s = 1, 2\}$ necessary for nonparametric combination are:

1. All permutation partial test T_{j_s} are marginally unbiased and significant for large values, so that they are stochastically larger in H_1 than in H_0.
2. All permutation partial tests T_{j_s} are consistent, that is,

$$\Pr\{T_{j_s} \geq T_{j_s\alpha}|H_{1j_s}\} \rightarrow 1, \; \forall\alpha > 0, j = 1, \ldots, K, s = 1, 2,$$

   as n tends to infinity, where $T_{j_s\alpha} < +\infty$ is the critical value of T_{j_s} at level α. In order to obtain global traditional consistency it suffices that at least one partial test is consistent (Pesarin 2001; Pesarin and Salmaso 2010b).

Let $\lambda_{j_s}, j = 1, \ldots, K, s = 1, 2$ be the set of p-values associated with partial tests in $\mathbf{T}$, that are positively dependent in the alternative and this irrespective of dependence relations among component variables in V.

In shape analysis field, $j = 1, \ldots, K, s = 1, 2$ represents the K landmarks in two dimensions. In order to apply NPC methodology, usually the hypothesis testing problem is broken down into two stages, considering both the coordinate and the landmark level (and, if present, the domain level too). Hence, we formulate partial test statistics for one-sided hypotheses and then we consider the global test T'' obtained after combining at the first stage with respect to s, then with respect to j (of course, this sequence may be reversed).

We wish to remark that in Brombin (2009); Brombin and Salmaso (2013), the effect of Generalized Procrustes Analysis (GPA) superimposition on the power of NonParametric Combination (NPC) tests has been investigated throughout a simulation study. Actually, it has been shown that including GPA, NPC tests are approximate, since GPA superimposition provides permutationally non-equivalent transformations (Brombin 2009). Moreover, the probability distribution of transformed data after GPA may be altered with respect to the initial distribution. Hence GPA privileges the shape, but it may alter the dependency structures and, as a result, the distribution producing permutationally non-equivalent tests within the permutation testing framework In the extreme case, if we consider two shapes that differ only for a scale factor (e.g. a big and a small circle), without GPA, inferential results obtained using NPC tests lead us to accept the alternative hypothesis, i.e. the two shapes are significantly different. On the other hand, after superimposition, we just accept the null hypothesis, stating the equality of the two shapes. Hence, inferential conclusions may be highly different.

4.2.3 A Suitable Algorithm

We now illustrate the algorithm for calculating the multivariate test, in its simplest version. Then we may add a multi-aspect procedure and adjust partial p-values for multiplicity through closed testing procedure (Finos and Salmaso 2007; Brombin 2009; Brombin and Salmaso 2009).

- The first phase (*coordinate level*) of a procedure estimates the distribution of $\mathbf{T}$ including the following steps:

1a. Calculate the vector of observed values of tests $\mathbf{T} : \mathbf{T}_o = \mathbf{T}(\mathbf{V}_{j_s})$.

1b. Consider a member g^*, randomly drawn from the set $\mathbf{G}$ of all possible permutations, and the values of vector statistics $\mathbf{T}^* = \mathbf{T}(\mathbf{V}^*_{j_s})$, where $\mathbf{V}^*_{j_s} = g^*(\mathbf{V}_{j_s})$. In most situations, the data permutation $\mathbf{V}^*_{j_s}$ may be obtained at first by considering a random permutation $(a^*_1, \ldots, a^*_n)$ of integers $(1, \ldots, n)$ and then by assignment of related individual data vectors to the proper group; thus, according to the unit-by-unit representation, $\mathbf{V}^*_{j_s} = \{\mathbf{V}_{j_s}(a^*_i), i = 1, \ldots, n; n_1, n_2\}$.

1c. Carry out B independent repetitions of step (b). The set of Conditional Monte Carlo (CMC) sampling results $\{\mathbf{T}^*_r, r = 1, \ldots, B\}$ is thus a random sampling from the permutation q-variate distribution of vector test statistics $\mathbf{T}$.

1d. The q-variate EDF $\hat{F}_B(\mathbf{z}|\mathbf{V}_{j_s}) = \left[\frac{1}{2} + \sum_r \mathbf{I}(\mathbf{T}^*_r \leq \mathbf{z})\right]/(B+1)$, $\forall \mathbf{z} \in \mathscr{R}^q$, gives an estimate of the corresponding q-dimensional permutation distribution $F(\mathbf{z}|\mathbf{V}_{j_s})$ di $\mathbf{T}$. Moreover,

$$\hat{L}_{j_s}(z|\mathbf{V}_{j_s}) = \left[\frac{1}{2} + \sum_r \mathbf{I}(\mathbf{T}^*_{j_s r} \geq z)\right]/(B+1), j = 1, \ldots, K, s = 1, \ldots, d,$$

gives an estimate $\forall \mathbf{z} \in \mathscr{R}^1$ of the marginal permutation significance level functions $L_{j_s}(z|\mathbf{V}_{j_s}) = \Pr\{T^*_{j_s} \geq z|\mathbf{V}_{j_s}\}$; this $\hat{L}_{j_s}(T_{j_s o}|\mathbf{V}_{j_s}) = \lambda_{j_s}$. This gives an estimate of the marginal p-value related to test T_{j_s}.

At the end of this first phase, we get a p-value for each landmark coordinate, hence in total Km, partial p-values.

If, for example, we analyze an object characterized by $K = 4$ landmarks in the plane ($m = 2$), hence $\lambda^*{}_1$ is the permutation p-value corresponding to the first tangent coordinate representing the position of the first landmark in the x direction, $\lambda^*{}_2$ the permutation p-value corresponding to the second tangent coordinate representing the position of the first landmark in the y direction, $\lambda^*{}_3$ the permutation p-value corresponding to the third tangent coordinate representing the position of the second landmark in the x direction, $\lambda^*{}_4$ is the permutation p-value corresponding to the fourth tangent coordinate representing the position of the second landmark in the y direction and so on (see Fig. 4.1).

Fig. 4.1 Algorithm for $K = 4$ two-dimensional landmarks and two domains combinations

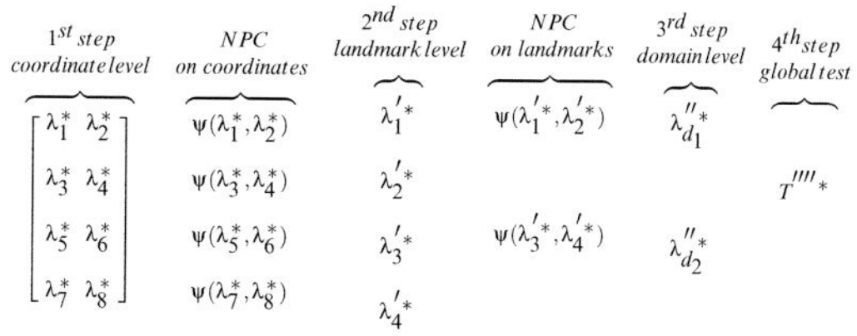

- The second phase (*landmark level*) of the algorithm includes the following steps.

2a. The q observed p-values are estimated from the data $\mathbf{V}_{j_s}$ by $\lambda_{j_s} = \hat{L}_{j_s}(T_{j_s o}|\mathbf{V}_{j_s})$, where $T_{j_s o} = T_{j_s}(\mathbf{V}_{j_s})$, $j = 1, \ldots, K$, $s = 1, \ldots, m$, represent the observed values of partial tests and $\hat{L}_{j_s}$ is the j_sth marginal significance level function, the latter being jointly estimated by the Conditional Monte Carlo (CMC) sampling method on data set $\mathbf{V}_{j_s}$, in accordance with step (1d) above.

2b. The combined observed value of the second-order test is evaluated through the same CMC results of the first phase, and is given by the combination of sequential couples (or triplets) of landmark indexes (landmark coordinates) as illustrated in Fig. 4.1. For example the observed statistic related to the first landmark (in two-dimensional case), is given by

$$T''_{1o} = \psi(\lambda_1, \lambda_2).$$

2c. The rth combined value of vector statistics (step (1d)) for the first landmark is then calculated by

$$T''^{*}_{1r} = \psi(\lambda^*_{1r}, \lambda^*_{2r}),$$

where $\lambda^*_{1r} = \hat{L}_1(T^*_{1r}|\mathbf{V}_{j_s})$, $r = 1, \ldots, B$.
Steps (2b) and (2c) will be repeated K times, in order to obtain a partial p-value for each landmark

- The third phase (*domain level*) of the algorithm include the following steps.

3a. Let us assume that Z out of K landmarks, $1 \leq Z \leq K$, constitute the first domain (i.e. a subgroup of landmarks sharing anatomical, biological or locational features); A out of K landmarks, $1 \leq A \leq K$, constitute the second domain and C out of K landmarks, $1 \leq C \leq K$, constitute the third domain. We have just defined three domains but, of course, we may define more than three domains.

3b. The combined observed value of the third-order test is evaluated through the same CMC results of the second phase, and is given by

$$T'''_{Zo} = \psi(\lambda'_1, \ldots, \lambda'_Z).$$

corresponding to the first domain,

$$T'''_{Ao} = \psi(\lambda'_1, \ldots, \lambda'_A).$$

corresponding to the second domain, and

$$T'''_{Co} = \psi(\lambda'_1, \ldots, \lambda'_C).$$

corresponding to the third domain.

3c. The rth combined value of vector statistics is then calculated by

$$T'''^{*}_{Zr} = \psi(\lambda'^{*}_{1r}, \ldots, \lambda'^{*}_{Zr}),$$

where $\lambda'^{*}_{zr} = \hat{L}_z(T'''^{*}_{zr}|\mathbf{V})$, $z = 1, \ldots, z$, $r = 1, \ldots, B$, is the permutation p-value corresponding to landmarks belonging to the first domain;

$$T'''^{*}_{Ar} = \psi(\lambda'^{*}_{1r}, \ldots, \lambda'^{*}_{Ar}),$$

where $\lambda^{*}_{ar} = \hat{L}_a(T'''^{*}_{ar}|\mathbf{V})$, $a = 1, \ldots, A$, $r = 1, \ldots, B$, is the permutation p-value corresponding to landmarks belonging to the second domain;

$$T'''^{*}_{Cr} = \psi(\lambda'^{*}_{1r}, \ldots, \lambda'^{*}_{Cr}),$$

where $\lambda'^{*}_{cr} = \hat{L}_c(T'''^{*}_{cr}|\mathbf{V})$, $c = 1, \ldots, C$, $r = 1, \ldots, B$, is the permutation p-value corresponding landmarks belonging to the third domain;

Hence at the end of this step we obtain different p-values corresponding to predefined domains. Figure 4.1 illustrates an example where we have defined 2 domains, namely d_1 and d_2, combining landmarks $1, 2$ and landmarks $3, 4$ respectively.

- The fourth and last phase provides the global p-value.

4a. The combined observed value of the global test is evaluated through the same CMC results in the first phase, and is given by:

$$T''''_o = \psi(\lambda'^{*}_1, \lambda'^{*}_2, \lambda''^{*}_Z, \ldots, \lambda''^{*}_A, \ldots, \lambda''^{*}_C).$$

4b. The rth combined value of vector statistics (step (S.d$_k$)) is then calculated by

$$T''''^{*}_r = \psi(\lambda'^{*}_{1r}, \lambda'^{*}_{2r}, \lambda''^{*}_{Zr}, \ldots, \lambda''^{*}_{Ar}, \ldots, \lambda''^{*}_{Cr}).$$

4c. Hence, the p-value of the combined test T'''' is estimated as

$$\lambda''''_{\psi} = \sum_r \mathbf{I}(T''''^{*}_r \geq T''''_o)/B.$$

4d. If $\lambda''''_{\psi} \leq \alpha$, the global null hypothesis H_0 is rejected at significance level α.

This algorithm may be generalized to include multi-aspect (MA) evaluations (Pesarin and Salmaso 2010b; Brombin and Salmaso 2013). As well known, different tests of significance are appropriate to test different features of the same null hypothesis (Fisher 1935). Actually a certain treatment or environment may affect/influence not only location but also scale coefficients or other aspects: these hypotheses may conveniently be examined through several test statistics, each one sensitive to differences that affect a particular aspect of the two distributions.

To summarize, the MA procedure embodies three steps: combination.

- definition of the aspects of interest and selection of a suitable test statistic for each aspect;
- organization of the aspects in a hierarchical structure;
- choice of a proper combining function to combine within and between aspects.

MA approach aims to supply a global evaluation on the basis set of partial tests, allowing also for the vice versa. Partial and global tests are exact, unbiased and consistent and MA is robust under very mild conditions (Salmaso and Solari 2005).

One of the main feature and advantage of the proposed approach is that using the MA procedure and the information about domains we are able to obtain not only a global p-value, like in traditional tests, but also a p-value for each of the defined aspects or domains. Hence following our procedure it is possible to construct a hierarchical tree, allowing for testing at different levels of the tree. On one hand partial tests may provide marginal information for each specific aspect, on the other they jointly provide information on the global hypothesis. In this way, if we find a significant departure from H_0, we can investigate the nature of this departure in detail. Also, one can move from the top to the bottom of the tree and, for interpreting results in a hierarchical way, from the bottom to the top. It is worth noting that "intermediate" level p-values need to be adjusted for multiplicity.

4.3 General Framework for Longitudinal Data Analysis in NPC Framework

Let us now assume that landmark data are available on different individuals at a common set of times, taking the form of a 5-way array, $\{X^{\dagger}_{g_{ij_s}}(t)\}$, where again $i = 1, \ldots, N$ labels different individuals, $g = 1, \cdots, G$ labels the groups in which subject i belongs to, $j = 1, \ldots, K$ labels different landmarks, $s = 1, \ldots, m$ indicates different dimensions and $t = 1, \ldots, T$ labels different times. As seen in previous sections, it is possible to represent these data as a collection $\{X^{\dagger}_{g_i}(t)\}$ of $K \times m$ matrices. As well known, the direct analysis of databases of landmark locations is not convenient because of the presence of nuisance parameters, such as position, orientation and size. Generalized Procrustes Analysis (GPA) is usually performed to eliminate non-shape variation in configurations of landmarks and to align the specimens to a common coordinate system (Rohlf and Slice 1990). In light of these considerations, we use Procrustes tangent coordinates about a centered and scaled mean configuration $\boldsymbol{\mu}$, corresponding to the Generalised Procrustes estimate (Dryden and Mardia 1998) based on all NT configurations, but the exact choice does not matter (Kent et al. 2001).

For the purposes of this work, we shall ignore any differences between the individual subjects. Furthermore, we also ignore changes in size and limit attention only to changes in the shape of objects represented in the plane (i.e. we assume $m = 2$).

We assume that the response variables behave according to the following model:

$$V_{gij_s}(t) = \mu_{j_s} + \mu_{ij_s}(t) + \delta_{gj_s}(t) + \sigma_{j_s}(t)(\delta_{gj_s}(t))\, Z_{gij_s}(t) \tag{4.1}$$

where

- μ_{j_s} represents a population constant for the j_sth variable (i.e. landmark coordinate);
- $\mu_{ij_s}(t)$ represents a time effect on the j_sth variable at time t and specific to the ith individual;
- $\delta_{gj_s}(t)$ represents a group-time effect at level g on the j_sth variable;
- $\sigma_{j_s}(t)(\delta_{gj_s}(t)) > 0$ represent population scale coefficients for variable j_s at time t, which are assumed to be invariant with respect to units but which may depend on group levels through the effects $\delta_{gj_s}(t)$, provided that, when $\delta_{gj_s}(t) \neq 0$, stochastic dominance relationships $\{V_{jj_s}(t)\} \overset{d}{<} (\text{or} \overset{d}{>}) \{V_{rj_s}(t)\}$, $j \neq r = 1, \ldots, G$, are satisfied;
- $Z_{gij_s}(t)$ are the error terms of a q-variate random vector, $\mathbf{Z}$, which are assumed to be exchangeable with respect to treatment levels, independent with respect to units, with zero mean, $E(\mathbf{Z}) = \mathbf{0}$, and with unknown distribution $P \in \mathscr{P}$. In particular, these errors may be temporally correlated and the temporal dependence studied through any kind of monotonic regression (Pesarin and Salmaso 2010b).

In this setting, different hypotheses may be of interest. Actually it is possible to evaluate separately group effect and time effects, or to jointly evaluate changes among groups and throughout times. In the first case, a *time-to-time* analysis is performed and the problem reduces to perform a series of one-way MANOVA. In the second case, the problem reduces to perform a series of tests for paired samples to compare, within each group, times or, if appropriate/sensible, it may be solved within a stochastic ordering framework (Basso and Salmaso 2011). Finally, the latter case may be solved performing a two-way MANOVA and applying the synchronized permutations as proposed in Basso et al. (2009).

References

Alfieri R, Bonnini S, Brombin C, Castoro C, Salmaso L (2012) Iterated combination-based paired permutation tests to determine shape effects of chemotherapy in patients with esophageal cancer. Stat Methods Med Res. doi:101177/0962280212461981. Article published online before print

Amaral G, Dryden I, Wood A (2007) Pivotal bootstrap methods for k-sample problems in directional statistics and shape analysis. J Am Stat Assoc 102:695–707

Basso D, Salmaso L (2011) A permutation test for umbrella alternatives. Stat Comput 21:45–54

Basso D, Pesarin F, Salmaso L, Solari A (2009) Permutation tests for stochastic ordering and ANOVA: theory and applications in R. Springer, New York

Birnbaum A (1962) On the foundations of statistical inference. J Am Stat Assoc 57:269–326

Blair RC, Higgins JJ, Karniski W, Kromrey JD (1994) A study of multivariate permutation tests which may replace Hotelling's t^2 test in prescribed circumstances. Multivar Behav Res 29:141–163

Bookstein FL (1997) Shape and the information in medical images: A decade of the morphometric synthesis. Comput Vis Image Underst 66:97–118

Brombin C (2009) A nonparametric permutation approach to statistical shape analysis. Ph.D. thesis. University of Padova, Padova, Italy

Brombin C, Salmaso L (2009) Multi-aspect permutation tests in shape analysis with small sample size. Comput Stat Data Anal 53:3921–3931

Brombin C, Salmaso L (2013) Permutation tests for shape analysis. Springer briefs in statistics. Springer, New York

Brombin C, Pesarin F, Salmaso L (2008) Dealing with more variables than sample sizes: an application to shape analysis. In: Hunter DR, Richards DSP, Rosenberger JL (eds) Nonparametric statistics and mixture models: a festschrift in honor of Thomas P. Hettmansperger. World Scientific, Singapore, pp 28–44

Brombin C, Mo G, Zotti A, Giurisato M, Salmaso L, Cozzi B (2009a) A landmark analysis-based approach to age and sex classification of the skull of the mediterranean monk seal (monachus monachus) (hermann, 1779). Anat Histol Embryol 38:382–386

Brombin C, Salmaso L, Villanova C (2009b) Multivariate permutation shape analysis with application to aortic valve morphology. In: Capasso V et al (eds) Stereology and image analysis. Ecs10: proceeding of the 10th european conference of ISS, The MIRIAM project series, vol 4. Esculapio Publishing Co., Bologna, Italy, pp 442–449

Celant G, Pesarin F, Salmaso L (2000) Two sample permutation tests for repeated measures with missing values. J Appl Stat Sci 9:291–304

Dryden IL, Mardia KV (1993) Multivariate shape analysis. Sankhyā Ser A 55:460–480

Dryden IL, Mardia KV (1998) Statistical shape analysis. Wiley, London

Dryden IL, Kume A, Le H, Wood ATA (2008) A multi-dimensional scaling approach to shape analysis. Biometrika 95(4):779–798

Edgington E, Onghena P (2007) Randomization tests, 4th edn. Chapman and Hall/CRC, London

Edwards A (1972) Likelihood. Cambridge University Press, Cambridge

Finos L, Salmaso L (2007) FDR- and FWE-controlling methods using data-driven weights. J Stat Plan Inference 137:3859–3870

Fisher RA (1935) The design of experiments. Oliver & Boyd, Edinburgh

Good P (2000) Permutation tests: a practical guide to resampling methods for testing hypotheses. Springer, New York

Good P (2005) Permutation, parametric, and bootstrap tests of hypotheses, 3rd edn. Springer, New York

Kent J, Mardia K, Morris R, Aykroyd R (2001) Functional models of growth for landmark data. In: Mardia K, Kent J, Aykroyd R (eds) Proceedings in functional and spatial data analysis. Springer, Berlin, pp 75–86

Lele S, Cole TM (1996) A new test for shape differences when variance-covariance matrices are unequal. J Hum Evol 31:193–212
Pesarin F (2001) Multivariate Permutation tests: with application in biostatistics. Wiley, Chichester, NY
Pesarin F (2002) Extending permutation conditional inference to unconditional one. Stat Methods Appl 11:161–173
Pesarin F, Salmaso L (2010a) Finite-sample consistency of combination-based permutation tests with application to repeated measures designs. J Nonparametr Stat 22:669–684
Pesarin F, Salmaso L (2010b) Permutation tests for complex data: theory, applications and software. Wiley, New York
Pesarin F, Salmaso L (2012) A review and some new results on permutation testing for multivariate problems. Stat Comput 22:639–646
Preston S, Wood A (2010) Two-sample bootstrap hypothesis tests for three-dimensional labelled landmark data. Scand J Stat 37(1):568–587
Preston S, Wood A (2011) Bootstrap inference for mean reflection shape and size-and- shape from three-dimensional labelled landmark data. Biometrika 98(1):49–63
Rohlf FJ (2000) Statistical power comparisons among alternative morphometric methods. Am J Phys Anthropol 111:463–478
Rohlf FJ, Slice DE (1990) Extensions of the procrustes method for the optimal superimposition of landmarks. Syst Zool 39:40–59
Salmaso L, Solari A (2005) Multiple aspect testing for case-control designs. Metrika 12:1–10
Terriberry TB, Joshi S, Garig G (2005) Hypothesis testing with nonlinear shape models. In: Information processing in medical imaging, vol 3565. Springer Berlin, Heidelberg, pp 15–26

Chapter 5
Applications of NPC Methodology

Abstract In this Chapter, we show by means of a motivating example related to the analysis of the FG-NET database, that NonParametric Combination (NPC) tests can be effective tools when testing whether there is a difference between dynamics of facial expressions or testing which of the landmarks are more informative in explaining their dynamics. Moreover a NPC solution for assessing shape asymmetry in dynamic shape data is presented.

Keywords Matching and object symmetry • NPC test for longitudinal data • Paired landmark data • Testing object symmetry

5.1 Introduction

Modelling and carrying out inference on dynamic shapes is tricky and assessment of within-subject or between-subjects changes over time is not an easy task.

As discussed in Durrleman et al. (2013), dynamic shape analysis differs substantially from the usual cross-sectional analysis and, in general, no consensus has emerged about how to combine shape changes over time and shape changes across subjects. Hence, there is still no single approach which can be considered uniformly as being the most appropriate solution for these specific problems.

Most of the works proposed in the recent literature on dynamic shape analysis focuses on the description of the time-varying deformation of the ambient space in which the objects of interest lie. In many cases the interest is in proposing a model with merely descriptive purposes. That is, we look for a model which enables the description of a mean growth scenario representative of the population and the variations of this scenario both in terms of shape changes and in terms of change in growth speed. For a discussion on specific case studies we refer, for example, to Durrleman et al. (2013); Fontanella et al. (2013); Fishbaugh et al. (2012); Kume (2000); Kent et al. (2001), and Le and Kume (2000).

Without doubts, Linear Mixed Effects (LME) models (Laird and Ware 1982; Pinheiro and Bates 2000) provide a flexible and powerful statistical framework for the analysis of longitudinal data (see, for example, Fitzmaurice et al. 2011; Verbeke

C. Brombin et al., *Parametric and Nonparametric Inference for Statistical Dynamic Shape Analysis with Applications*, SpringerBriefs in Statistics,
DOI 10.1007/978-3-319-26311-3_5

and Molenberghs 2000) allowing to analyze simultaneously multiple longitudinal outcomes (Verbeke et al. 2010).

A joint model assumes a mixed model for each outcome such that univariate models are combined through the specification of a joint multivariate distribution for all random effects (Verbeke and Fieuws 2005). However, as the number of outcomes increases, the dimensionality of the random-effects covariance matrix becomes very large, thus leading to methodological and computational challenges. The computational complexity of high-dimensional joint random-effects models may be reduced by reducing the dimensionality of the problem.

Instead of estimating a full multivariate model, Barry and Bowman (2008) proposed an approach involving fitting bivariate LME models to all the pairwise combinations of individual tangent coordinates, and aggregating the results across repeated parameter estimates. The methodology yields unbiased estimates with robust standard errors reflecting the true sampling variability (Fieuws and Verbeke 2006).

Although appealing, the method is limited in the number of landmarks that may be evaluated due to the computational problems. Moreover, comparison of nested models is not straightforward. The Wald test, the likelihood ratio (LR) test and the pseudolikelihood ratio (PLR) test have been shown to perform poorly and have drawbacks when applied to either high-dimensional data or random effects models (Barry and Bowman 2008; Barry 2008). Their statistics may be influenced by directions of variation that are not important and this tends to get worse as the number of outcome variables increases. Moreover, PLR test evaluates whether all of the multiple copies of parameter estimates are equal to zero, rather than the individual parameters. The alternative bootstrap-based test proposed by (Faraway 1997), while overcoming some of the above-mentioned problems, is extremely time consuming, with both the pairwise full and submodels necessarily being fitted for each simulation.

The running example of this Chapter refers to the study of facial expressions. This is of much interest in medicine where facial expressions are related to the clinical manifestation of several neuropsychiatric disorders and various mental health problems, such as phobia, post-traumatic stress disorder, attention deficits, and schizophrenia. The analysis of facial expressions has also great relevance in Social Signal Processing (Vinciarelli et al. 2009) where of primary interest is the ability to express and recognize social signals produced during social interactions (e.g. agreement, politeness, empathy, friendliness, conflict, etc), coupled with the ability to manage them in order to act wisely in human relations (Pantic et al. 2011).

Given these foundations from medical and behavioural sciences, many researchers appraise facial expression analysis from a computer vision perspective and attempt to create automated computational models for facial expression classification. The problem we consider here is relatively simpler in that we limit our analysis in capturing the appearance changes that occur during facial expression formation in terms of the intensity where, for intensity, we mean the magnitude of the change shown by the landmarks used to synthesize the facial activity. This analysis can provide useful information on the dominant dynamic characteristics of the expressions and can thus be important for further classification purposes.

5.2 The Data

Suppose landmark data are available on different individuals at a common set of times, taking the form of a 5-way array, $\{X^{\dagger}_{gijs}(t)\}$, where $i = 1, \ldots, N$ labels different individuals, $g = 1, \cdots, G$ labels the groups in which subject i belongs to, $j = 1, \ldots, K$ labels different landmarks, $s = 1, \ldots, m$ labels different dimensions and $t = 1, \ldots, T$ labels different times.

Sometimes it is convenient to represent these data as a collection $\{X^{\dagger}_{g_i}(t)\}$ of $K \times m$ matrices. However, instead of using raw data, generalized Procrustes analysis is carried out to concentrate exclusively on differences in shapes and eliminate sources of non-shape variability.

After GPA alignment, Procrustes residuals are used as an approximation of the tangent coordinates.

For the purposes of the present analysis, we shall ignore any differences between the individual subjects. Furthermore, we also ignore changes in size and limit attention only to changes in the shape of objects represented in the plane (i.e. we assume $m = 2$).

With reference to the FG-NET database, as in Chapter 2 (Section 2.6), we only focus on happiness (H) and surprise (S) expressions. We consider video sequences gathered from 16 different individuals and summarize the expressions through a reduced set of 34 landmarks manually placed on the face of each subject. 7 equi-spatially frames have been considered as representative of the dynamic of the expression, from baseline to apex.

Facial landmark configuration has been already displayed in Chap. 2 (Sect. 2.6, see Fig. 2.1).

The dynamics of the two expressions are described by the changes in time of the landmark coordinates.

To make sure the observed variation in our data is sufficiently small, and the distribution of points in the tangent space may be used as a satisfactory approximation to their distribution in shape space, we have compared the Euclidean distances between all pairs of points in tangent space against their Procrustes distances in shape spaces. Although not shown here, the analysis highlights a strong linear relationship with a slope very close to 1 confirming that the tangent space can be satisfactorily used for these data.

The Procrustes residuals from this pole thus are approximate tangent coordinates and, for each expression, they also represent departures of each data shape from the neutral state.

Let $\{\mathbf{V}_{g_i}(t)\}$ denote the $K \times 2$ (centered not Helmertized) matrix of Procrustes tangent coordinates of the data $\{\mathbf{X}^{\dagger}_{g_i}(t)\}$. To provide an example of the dynamics shown by the two expressions, we assume all N individuals are i.i.d. and we take a sample average of the Procrustes coordinates to get averaged data. The dynamics of these data, for the two expressions, are shown in Fig. 5.1 where, for each landmark, the closed circle and the arrow represent the position at the initial and final times, respectively. The happiness expression (left) is mainly characterized by the

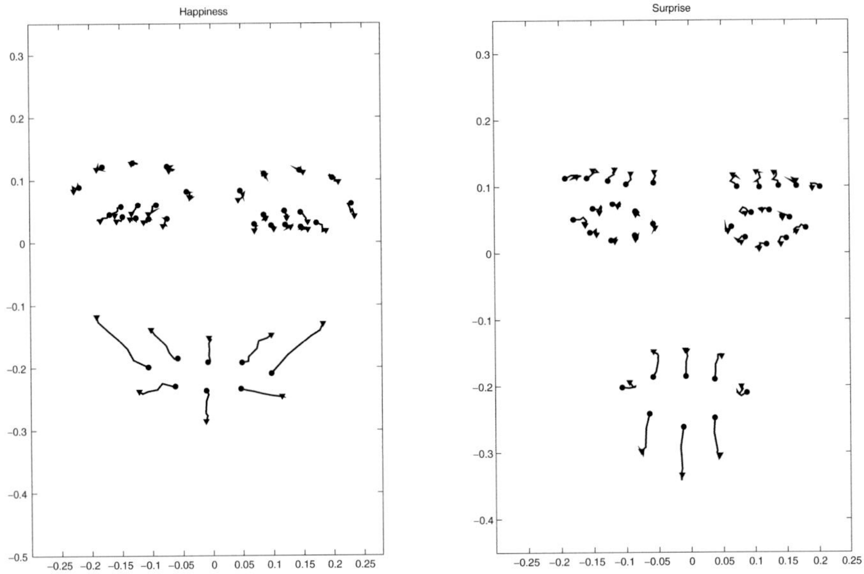

Fig. 5.1 Dynamics of averaged data for happiness and surprise expressions with patterns blown up by a factor of 3 for clarity. For each landmark, the *closed circle* and the *arrow* represent the position at the initial and final times, respectively

movements of the mouth since the eyes move only slightly. Specifically, we observe a slight narrowing of the eyelids and a raising of the lip corners describing an upward curving of mouth and expansion on vertical and horizontal direction. On the other hand, surprise (right) appears with a vertical expansion of the mouth, widened eyes and slightly raised eyelids and eyebrows. An asymmetry is also surprisingly observed between the movement of the upper eyelids for the two eyes.

5.3 NPC Methodology for Longitudinal Data

In this section we describe the NPC methodology dealing with paired and longitudinal data. In a general setting, it is assumed that a q-dimensional non-degenerate variable is observed at T different time occasions for N individuals in two experimental situations, corresponding to two groups.

With reference to our experimental study, there is thus complete information on $q = 68$ Procrustes tangent coordinates (i.e., K=34 landmarks in $m = 2$ dimensions) recorded at $T = 7$ different time occasions (i.e. frames), in a sample of $N = 16$ subjects performing both happy and surprise expressions.

We assume that the response variables behave according to the following model:

$$V_{gij_s}(t) = \mu_{j_s} + \mu_{ij_s}(t) + \delta_{gj_s}(t) + \sigma_{j_s}(t)(\delta_{gj_s}(t))Z_{gij_s}(t) \quad (5.1)$$

where

- $g = 1, 2$ and $g = 1$ denotes "Happiness" and $g = 2$ "Surprise";
- μ_{j_s} represents a population constant for the j_sth variable (i.e. landmark coordinate);
- $\mu_{ij_s}(t)$ represents a time effect on the j_sth variable at time t and specific to the ith individual;
- $\delta_{gj_s}(t)$ represents a group-time effect at level g on the j_sth variable which, without loss of generality, is assumed to be $\delta_{1j_s}(t) = 0$, $\delta_{2j_s}(t) \leq$ (or $\geq$) 0, $\forall (j_s, t)$;
- $\sigma_{j_s}(t)(\delta_{gj_s}(t)) > 0$ represent population scale coefficients for variable j_s at time t, which are assumed to be invariant with respect to units but which may depend on group levels (i.e. expressions) through the effects $\delta_{gj_s}(t)$, provided that, when $\delta_{2j_s}(t) \neq 0$, stochastic dominance relationships $\{V_{1j_s}(t)\} \overset{d}{<}$ (or $\overset{d}{>}$) $\{V_{2j_s}(t)\}$, are satisfied;
- $Z_{gij_s}(t)$ are the error terms of a q-variate (where $q = K \times m$ and in our application $q = K \times 2$ random vector, $\mathbf{Z}$, which are assumed to be exchangeable with respect to treatment levels, independent with respect to units, with zero mean, $E(\mathbf{Z}) = \mathbf{0}$, and with unknown distribution $P \in \mathscr{P}$. In particular, these errors may be temporally correlated and the temporal dependence studied through any kind of monotonic regression (Pesarin and Salmaso 2010).

Equation (5.1) provides a typical modelling framework for longitudinal data. However, it is worth noting that, provided the exchangeability assumption holds under H_0, other model formulations are possible—see, for example, Pesarin and Salmaso (2010).

Within the NPC approach, the data may be examined from various perspectives, focusing on different features: one could be interested in evaluating a group (expression) effect, a time effect or both. Hence the following hypothesis systems are of interest.

(a) By considering the two expressions, we perform a one-sample paired data analysis to test whether or not coordinates/landmarks, are equal in distribution at each time point. In short, we are interested in evaluating a group (expression) effect on coordinates/landmarks at each time (frame). At each time point, the q-dimensional vector of test statistics, $T^*_{k_s}$, is thus obtained as

$$T^*_{j_s}(t) = \sum_{i=1}^{N} \left[\bar{\mathbf{V}}_{1_i}(t) - \bar{\mathbf{V}}_{2_i}(t)\right] \cdot \mathscr{S}^*_i, \ i = 1, \ldots, N, \ t = 1, 2, \ldots, T;$$

where $\bar{\mathbf{V}}_{g_i}(t)$ is the *vec* of the $K \times 2$ matrix of Procrustes tangent coordinates $\mathbf{V}_{g_i}(t)$ and $\mathscr{S}^*_i$ are the $\pm$ signs, common to all variables, i.e. random realizations of the random variable $1 - 2\mathscr{B}n(1, 1/2)$, where $\mathscr{B}n(\cdot, \cdot)$ is the Binomial distribution.

Hence, the following partial hypotheses should be tested:

$$H_{0j_st}: \left\{V_{1j_s}(t) \overset{d}{=} V_{2j_s}(t)\right\},\ j_s = 1, \ldots, q,\ t = 1, 2, \ldots, T,$$

then suitably combined to test the null hypothesis:

$$H_{0t}: \left\{\bigcap_{j_s=1}^{q} H_{0j_st}\right\}, t = 1, 2, \ldots, T$$

$$H_{0j_s}: \left\{\bigcap_{t=1}^{T} H_{0j_st}\right\}, j_s = 1, 2, \ldots, q$$

against the partial alternatives

$$H_{1j_st}: \left\{V_{1j_s}(t) \overset{d}{<\neq>} V_{2j_s}(t)\right\},$$

suitably combined in a global test, in which at least one null hypothesis is not true

$$H_{1t}: \left\{\bigcup_{j_s=1}^{q} H_{1j_st}\right\},\ t = 1, 2, \ldots, T$$

$$H_{1j_s}: \left\{\bigcup_{t=1}^{T} H_{1j_st}\right\},\ j_s = 1, 2, \ldots, q.$$

To simplify the notation, we have discussed the procedure considering the whole set of coordinates. However, as the main interest generally lies in landmarks, once obtained partial tests for coordinates, one should combine the tests on single landmark coordinates. Hence, as an intermediate step, once should consider the combination w.r.t. the dimension m. The null and alternative hypothesis systems becomes slightly more complicated and are written as

$$H_{0kt}: \left\{V_{1j_1}(t) \overset{d}{=} V_{2j_1}(t)\right\} \bigcap \left\{V_{1j_2}(t) \overset{d}{=} V_{2j_2}(t)\right\},\ j = 1, \ldots, K$$

$$H_{1kt}: \left\{V_{1j_1}(t) \overset{d}{<\neq>} V_{2j_1}(t)\right\} \bigcup \left\{V_{1j_2}(t) \overset{d}{<\neq>} V_{2j_2}(t)\right\}.$$

The test statistics are in accordance with the rule-of-thumb that *large is significant* (see Pesarin and Salmaso (2010)).

Results from the analysis, carried out at landmark level, are shown in Figs. 5.2 and 5.3, where partial and global p-values are represented through a heat map.

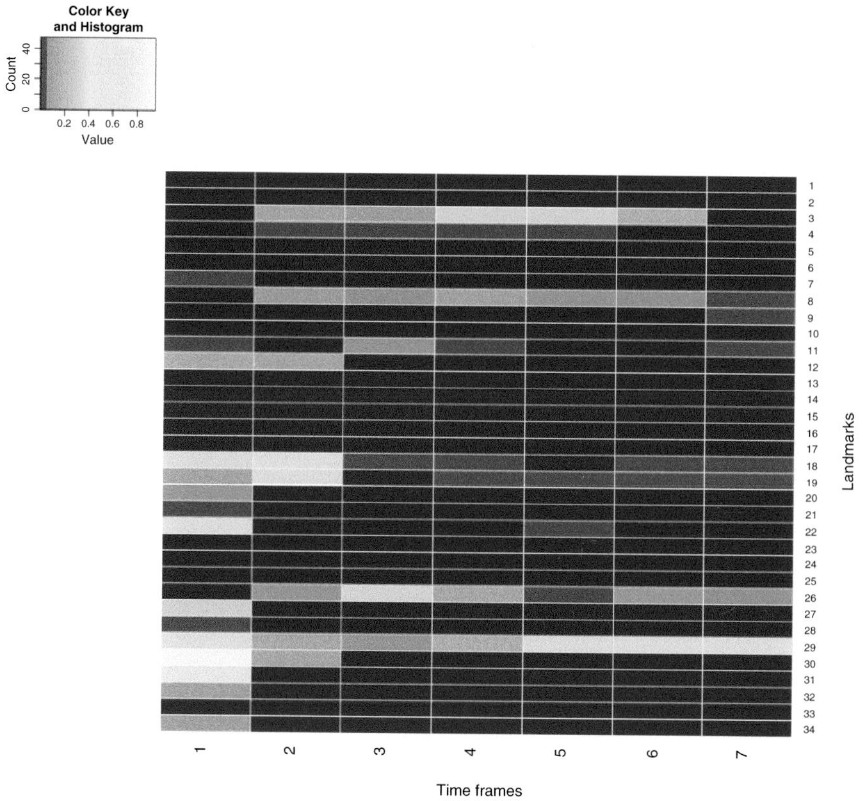

Fig. 5.2 Heat map representation of p-values. Evaluating whether happy and surprised facial expressions differ at landmark level within each time

The overall effect, after familywise error correction (FWE) and after combining throughout times, is instead displayed, for each landmark, in Fig. 5.4.

Figure 5.2 shows that the dynamics of the two expressions become increasingly evident with the progress of time. The difference between the expressions first appears in the eye regions (i.e. eyebrows and eyelids). At time-frame 1, for example, the p-values associated with the eye "domains" suggest that the differences between most of the landmarks are significant. In particular, at time-frame 1, all landmarks in the eyebrows are completely different between the expressions. Moreover, 75 % of the landmarks in the left eye and 62.5 % of the landmarks in the right eye are significantly different. This percentage increases at time-frame 3 up to 87.5 % and continues to increase until the end of the process (time-frame 7). On the other hand, at the beginning, only 25 % of the landmarks in the mouth region are significantly different, this percentage rapidly increases up to 75 % in the second time frame and continues to increase until the end of the process.

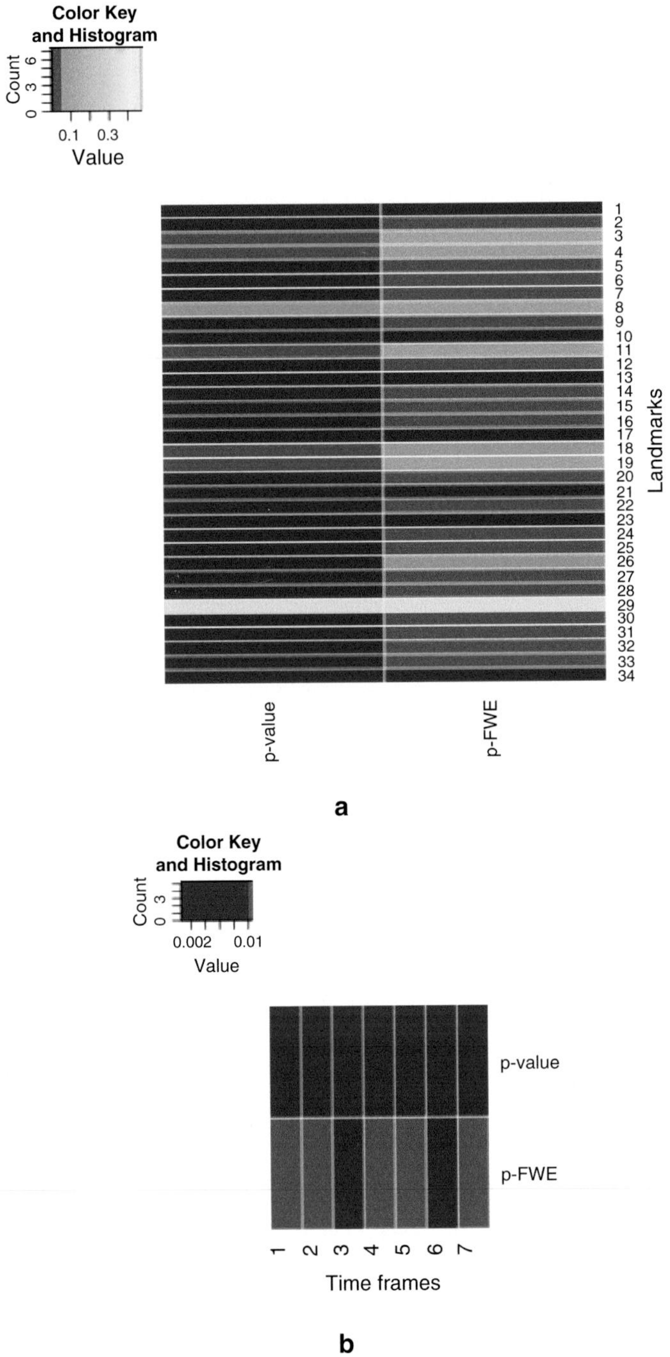

Fig. 5.3 Heat map representation of global *p*-values, obtained combining times (**a**) or landmarks (**b**)

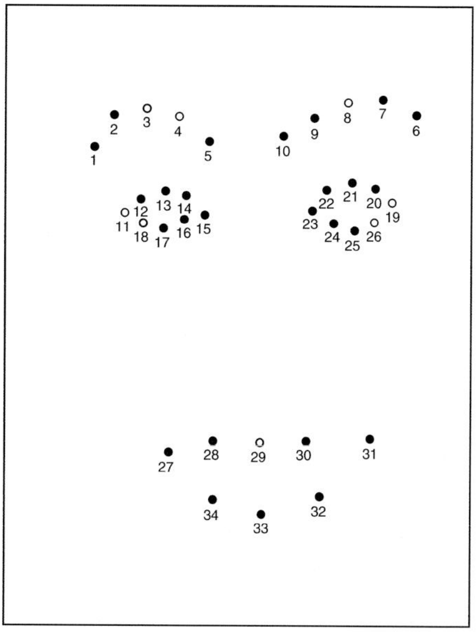

Fig. 5.4 Evaluating whether happy and surprised facial expressions are globally different at landmark level. Data points marked with a *solid circle* refer to landmarks with a significant p-value. Overall "group" effect for each landmark is shown, after combining throughout times and after FWE correction

At the end of the process, happy and surprise expressions appear completely different, with just two landmarks for which the test does not appear to be significant before correction for multiplicity (see Fig. 5.2, last column).

After FWE correction, most of the p-values obtained after combining times are significant with the exception of landmarks 3–4–8–11–18–19–26–29 (see Fig. 5.3a, right column, and Fig. 5.4). Hence, globally, happy and surprised expressions are significantly different as expected (see Fig. 5.3b).

Results from the same analysis carried out at coordinate levels are also shown in Figs. 5.5 and 5.6. In general, considering all the frames, the difference in the dynamics of the two expressions is mainly determined by the changes in the j_1 (horizontal) coordinates.

This finding is supported by Fig. 5.1 which shows that in the mouth region the dynamic of happy expression is mainly represented by a horizontal direction that seems to prevail changes in vertical direction observed in the surprised expression.

(b) It is of interest to evaluate the *time effect*.

For each expression, differences between consecutive times have been evaluated, thus obtaining the following test statistics: the q-dimensional vector of test statistics $T^*_{gj_s,r}$, $r = 1, 2 \ldots, T-1$ is obtained by considering the test statistics for the two expressions separately

$$T^*_{gj_s r} = \sum_{i=1}^{N} \left[\bar{\mathbf{V}}_{g_i}(r) - \bar{\mathbf{V}}_{g_i}(r+1)\right] \cdot \mathscr{S}^*_i, \; g = 1, 2, \; i = 1, \ldots, N.$$

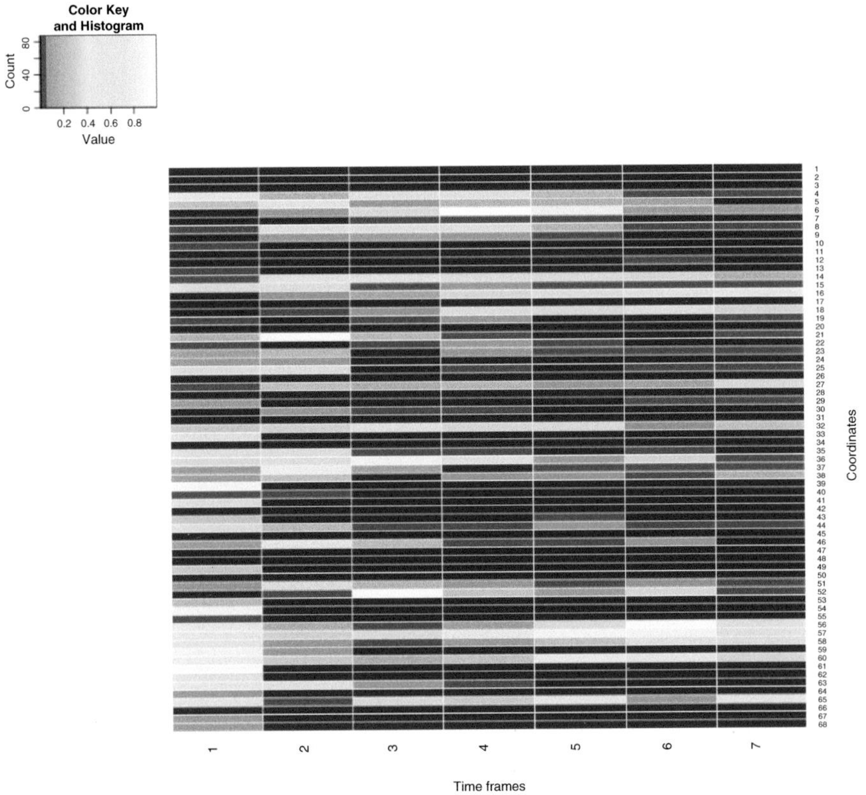

Fig. 5.5 Heat map representation of *p*-values. Evaluating whether happy and surprised facial expressions differ at coordinate level within each time

Hence, the following partial hypotheses should be tested:

$$H_{0gj_sr} : \left\{ V_{gj_s}(r) \stackrel{d}{=} V_{gj_s}(r+1) \right\},\ g = 1, 2,\ j_s = 1, \ldots, q,\ i = 1, \ldots, N,$$

then suitably combined to test the global null hypothesis:

$$H_{0g} : \left\{ \bigcap_{j_s=1}^{q} \left[\bigcap_{r=1}^{T-1} H_{0gj_sr} \right] \right\},\ g = 1, 2,$$

$$H_{0gj_s} : \left\{ \bigcap_{r=1}^{T-1} H_{0gj_sr} \right\},\ g = 1, 2,\ j_s = 1, \ldots, q,$$

against the partial alternatives

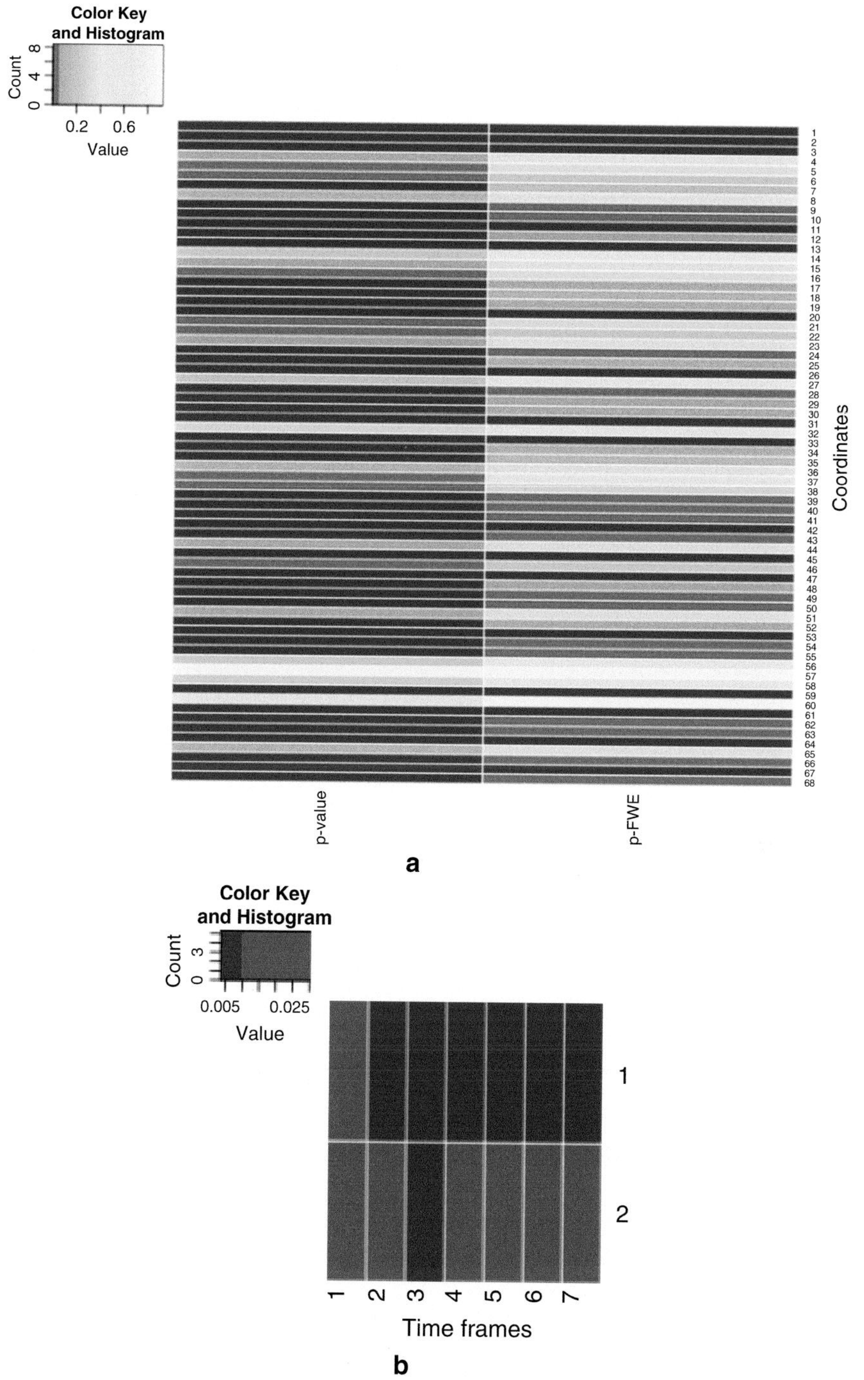

Fig. 5.6 Heat map representation of global *p*-values, obtained combining times (**a**) or coordinates (**b**)

$$H_{1gj_sr}: \left\{ V_{gij_s}(r) \overset{d}{<\neq>} V_{gij_s}(r+1) \right\},$$

suitably combined in a global test, in which at least one null hypothesis is not true

$$H_{1g}: \left\{ \bigcup_{j_s=1}^{q} \left[\bigcup_{r=1}^{T-1} H_{0gj_sr} \right] \right\}; \quad H_{1gj_s}: \left\{ \bigcup_{r=1}^{T-1} H_{0gj_sr} \right\}.$$

To simplify the notation, we have described the hypotheses at coordinate level. However, once obtained partial tests for coordinates, one should combine the tests on single landmark coordinates, thus obtaining a p-value for each landmark.

Results from the analysis carried out at landmark level for happiness and surprise are shown in Figs. 5.7, 5.8 and 5.9, 5.10, respectively, where partial and global p-values are represented through a heat map. The overall effects, after FWE correction,

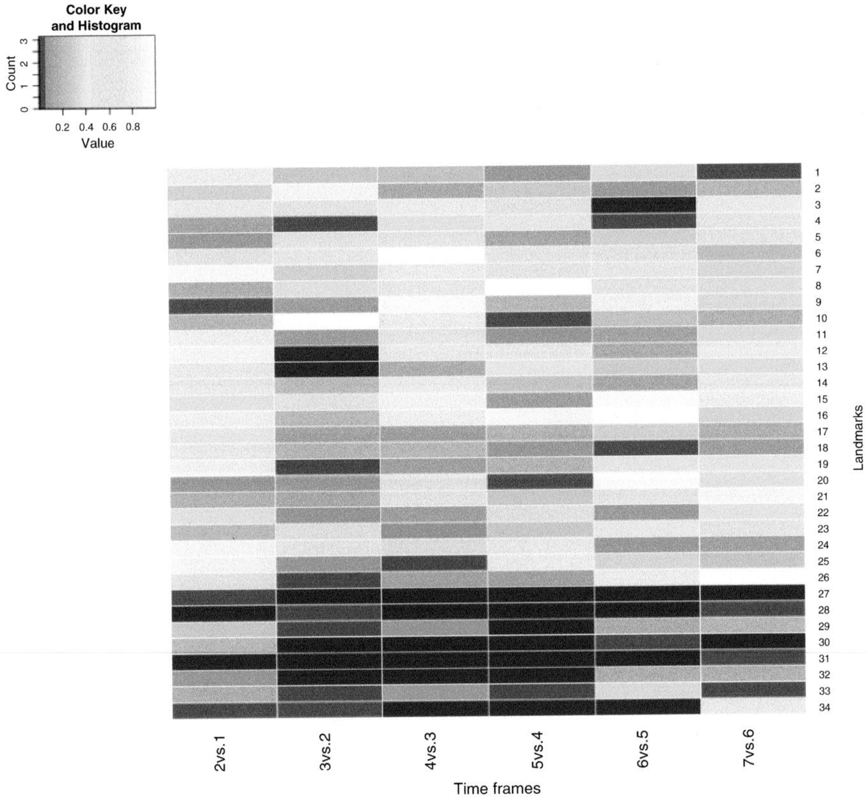

Fig. 5.7 Heat map representation of p-values. Evaluating time effect within happy facial expression at landmark level

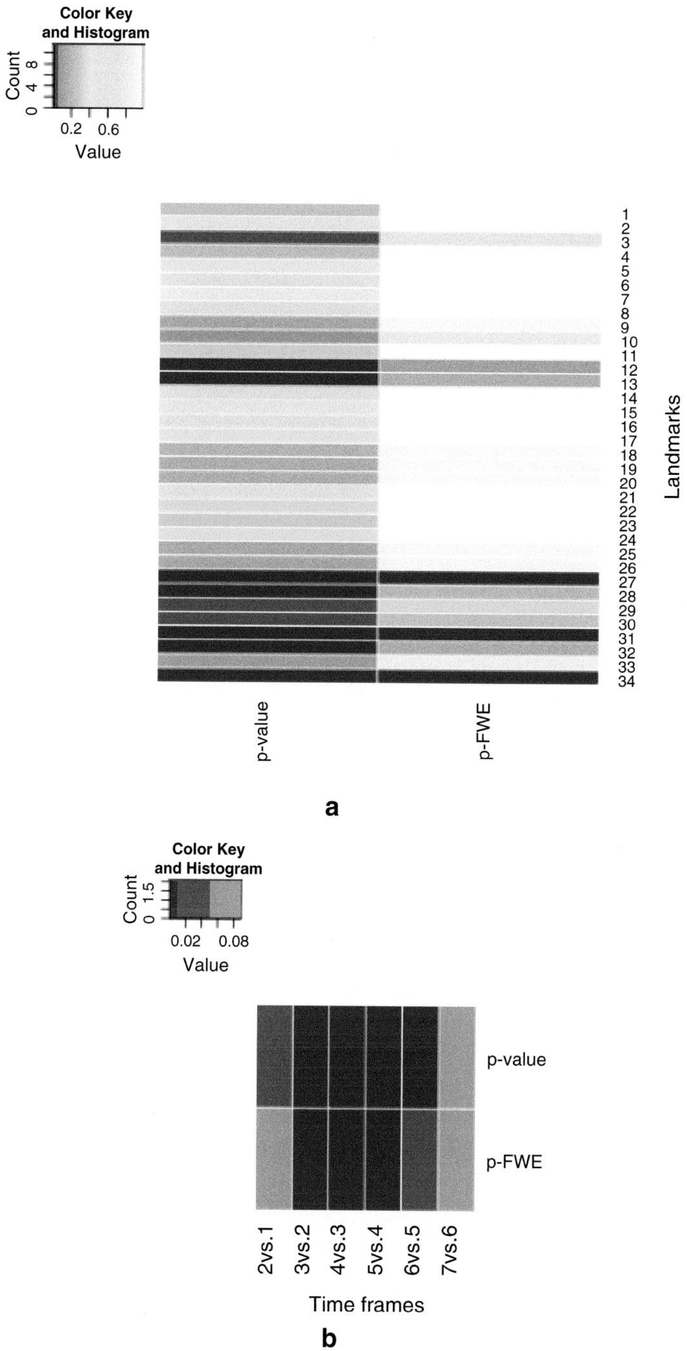

Fig. 5.8 Heat map representation of global p-values, obtained combining times (**a**) or landmarks (**b**)

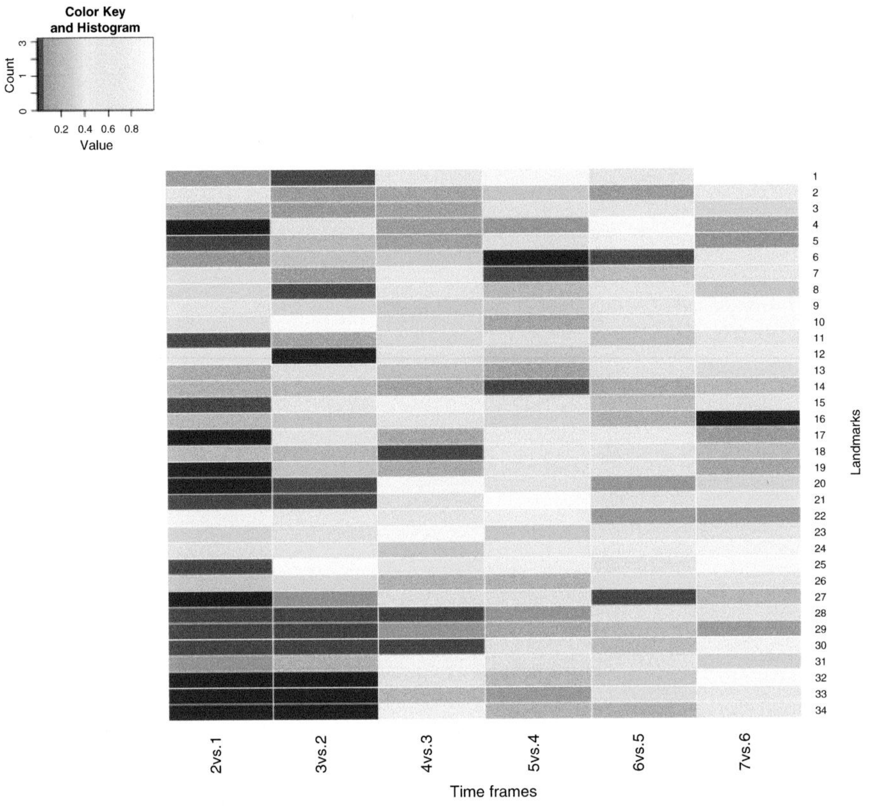

Fig. 5.9 Heat map representation of p-values. Evaluating time effect within surprised facial expression at landmark level

are also shown in Figs. 5.11 and 5.12. Results from the same analysis, carried out at coordinate levels are shown in Figs. 5.13, 5.14 and 5.15, 5.16.

In order to control the Familywise Error Rate and compute adjusted p-values, a Closed Testing Procedure has been applied (for details see Pesarin and Salmaso 2010).

With reference to happiness, we found a significant global time effect. After FWE correction, Fig. 5.8a, second column, and Fig. 5.11g suggest that most of the changes occur in the mouth region. Happiness is thus mainly characterized by the movements of the mouth, while moderate changes occur in he eye regions. Changes, for the most part, occur between time 2 and 3 (when comparing time frame 3 with the previous time frame 2), involving both mouth and eye regions.

Comparing the configuration at time 2 with the baseline, Fig. 5.7 (first column) highlights changes in the 50 % of the landmarks in the mouth region. When comparing times 2 and 3, this percentage grows up to 100 %. Moreover, most of the changes in the eye region (eyes and eyebrows) occurs between time frames 2

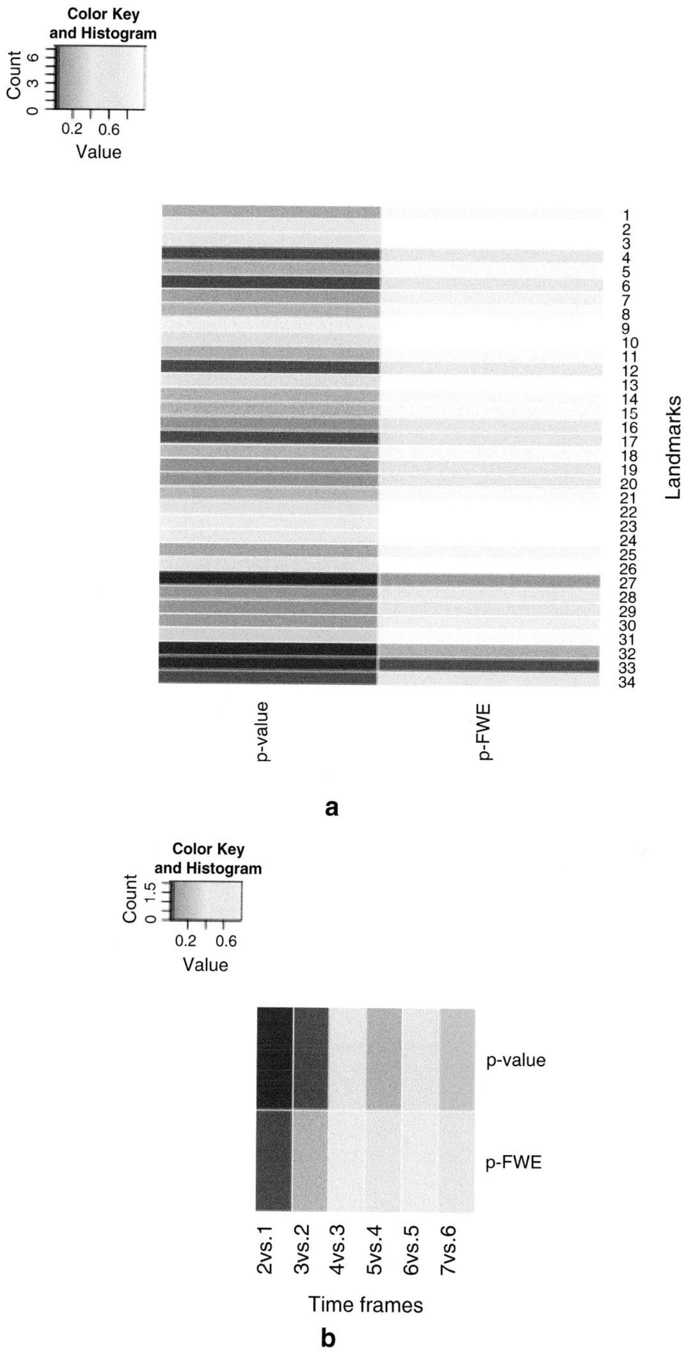

Fig. 5.10 Heat map representation of global *p*-values, obtained combining times (**a**) or landmarks (**b**)

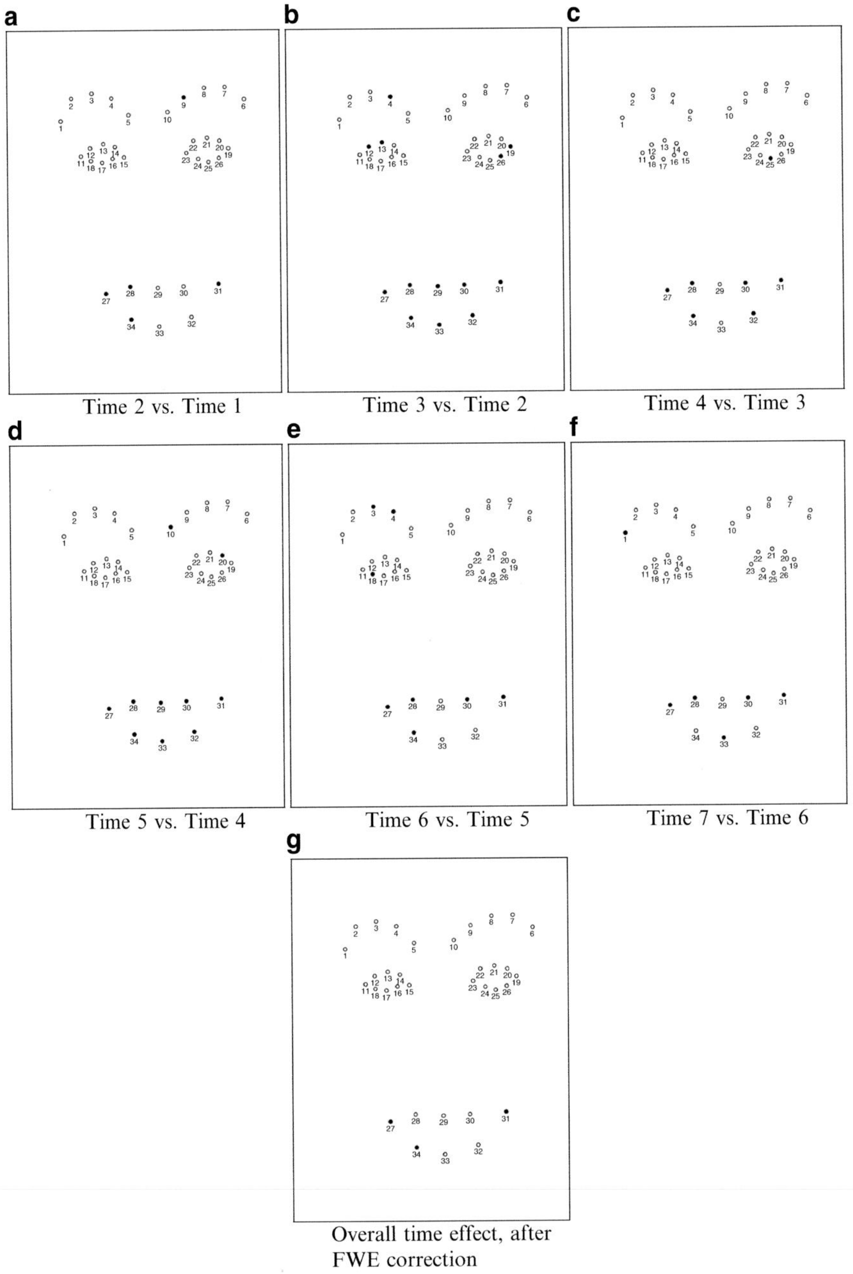

Fig. 5.11 Evaluating time effect within happy facial expression at landmark level. For each contrast, data points marked with a solid circle refer to landmarks with a significant *p*-value (**a**)–(**f**). Overall time effect, for each landmark, after FWE correction is shown in (**g**)

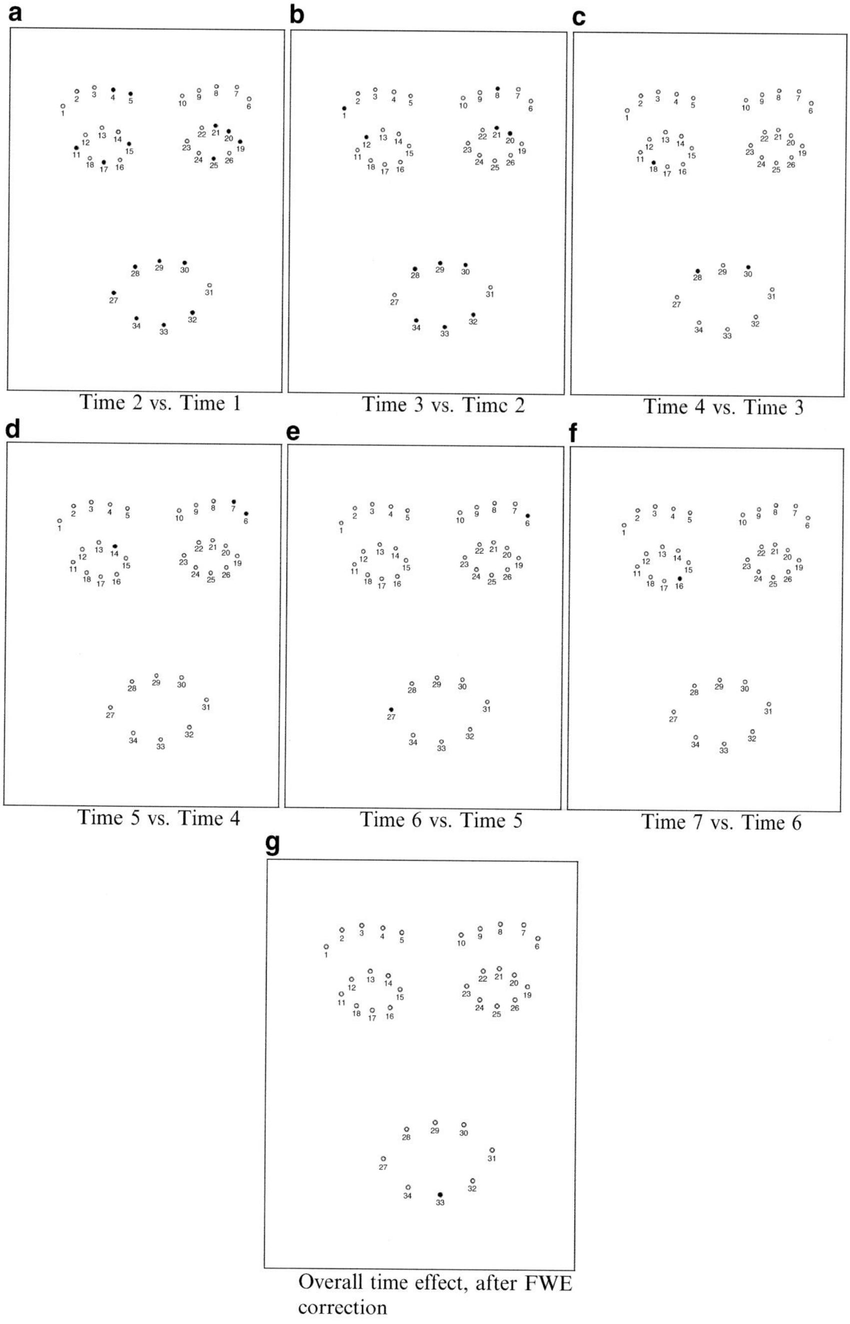

Fig. 5.12 Evaluating time effect within surprised facial expression at landmark level. For each contrast, data points marked with a solid circle refer to landmarks with a significant *p*-value (**a**)–(**f**). Overall time effect, for each landmark, after FWE correction is shown in (**g**)

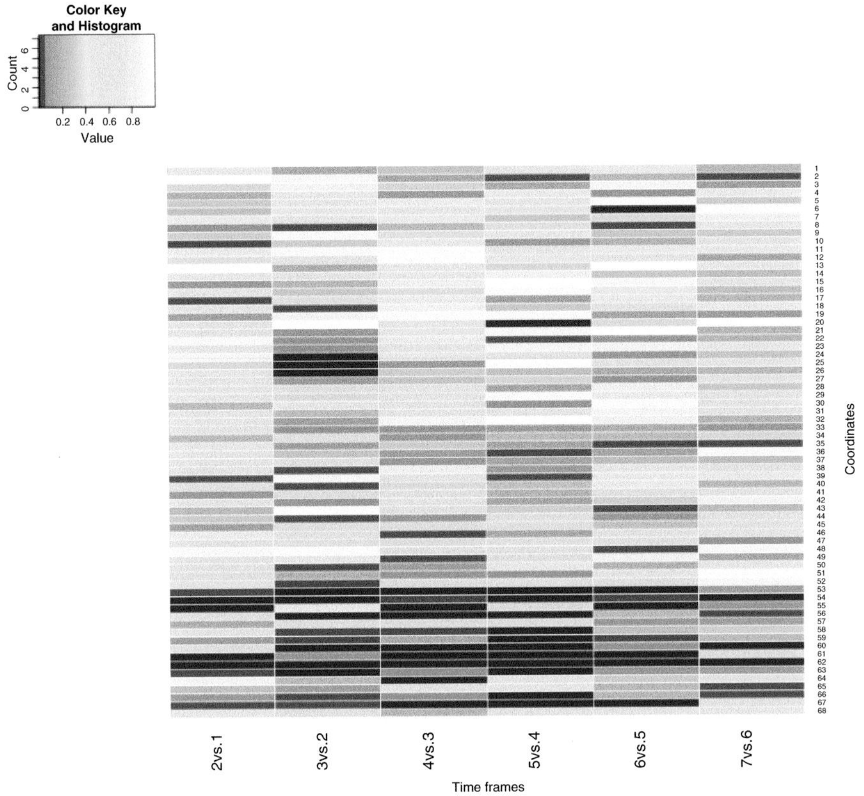

Fig. 5.13 Heat map representation of p-values. Evaluating time effect within happy facial expression at coordinate level

and 3, and, to a lesser extent, between times 5 and 6. In the facial configurations, landmarks with a significant p-values are represented as points marked with a solid circle.

Analyzing the heat map in Fig. 5.13, and considering all contrasts, we note that changes in the j_2 coordinates occur more frequently in the left and right eyebrows, in the left eye and in mouth region, while in the right eye horizontal changes slightly prevail. In general, at coordinate levels, vertical movements are more frequent than horizontal ones: more significant changes are observed in j_2 coordinates.

These results are in agreement with those found by applying polynomial regression models (see Sect. 3.4.1), where major involvement of mouth in the generation of happy facial expression was observed.

When considering surprised facial expression, after FWE correction, only one landmark is significant (see Figs. 5.10a, second column, and Fig. 5.12g).

When comparing the configuration at time 2 with the baseline (Fig. 5.9, first column), 87.5 % of the landmarks in the mouth region, 40 % of the landmarks in the

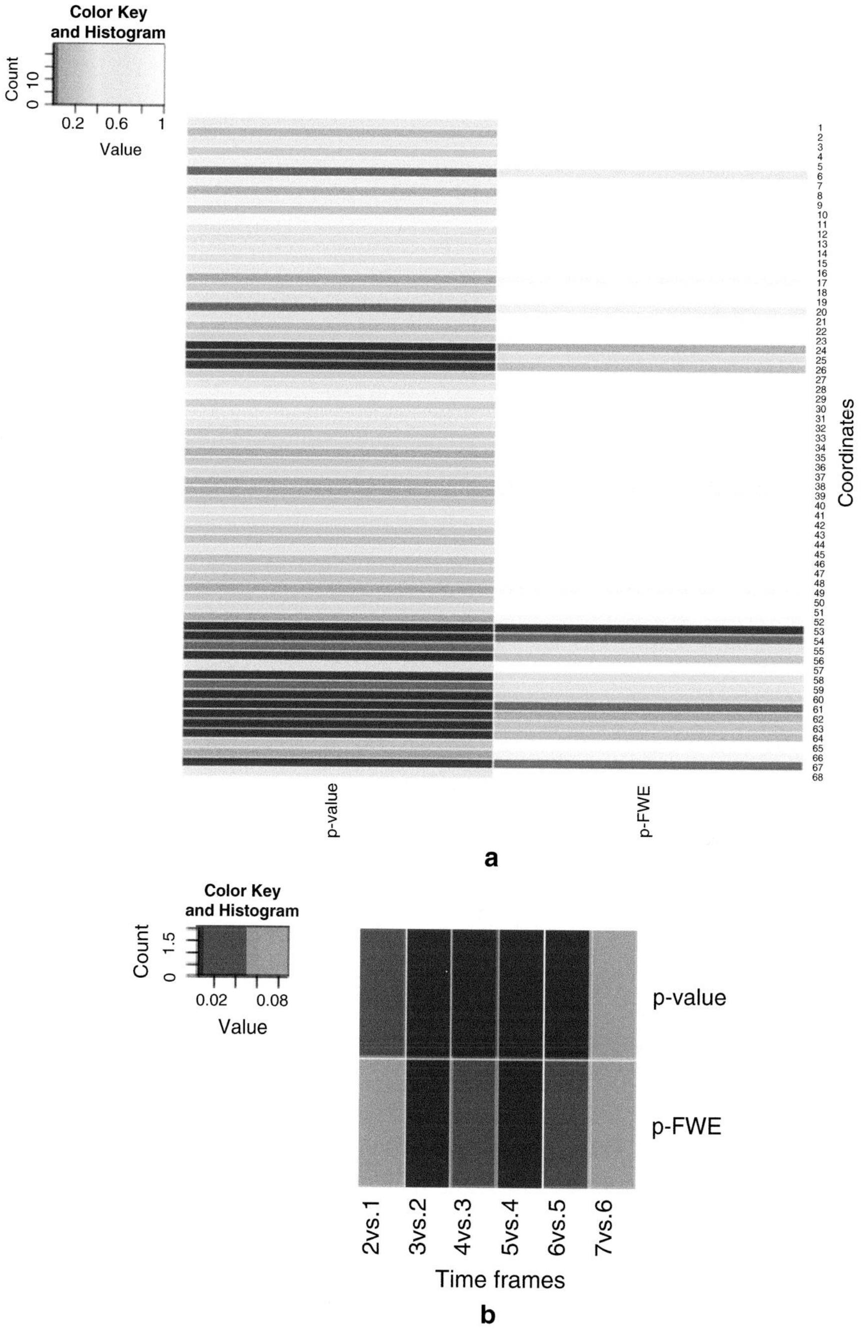

Fig. 5.14 Heat map representation of global p-values, obtained combining times (**a**) or coordinates (**b**)

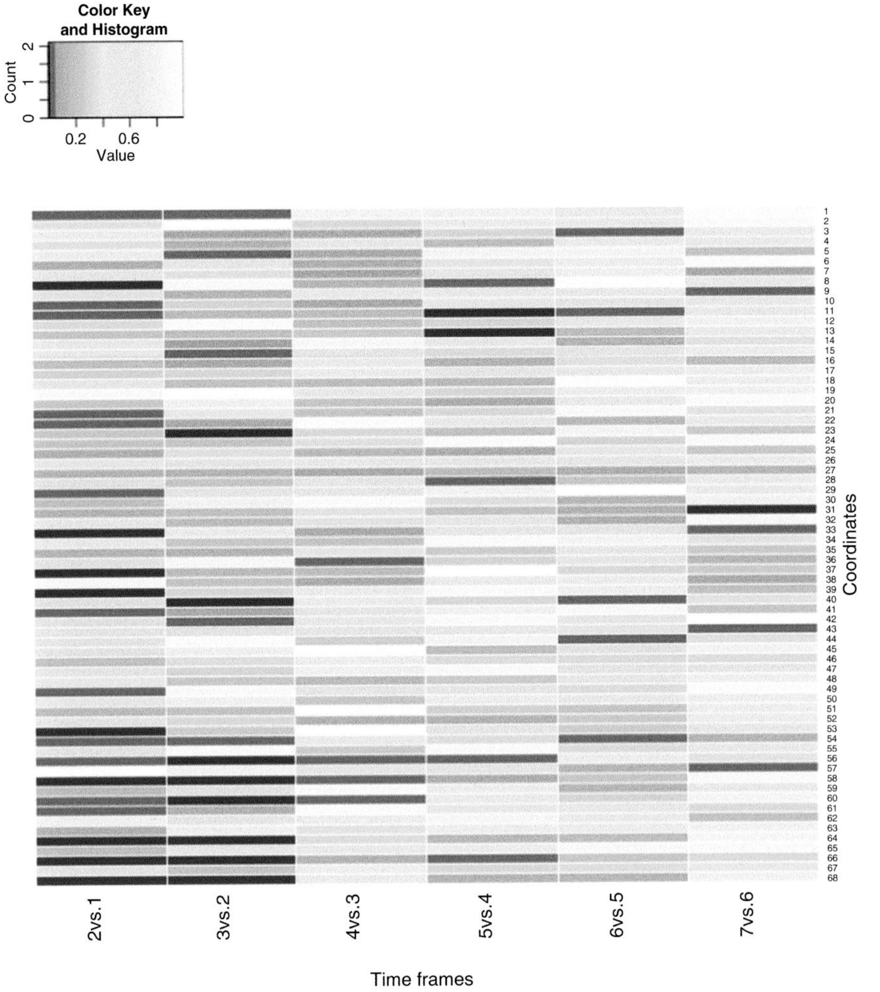

Fig. 5.15 Heat map representation of p-values. Evaluating time effect within surprised facial expression at coordinate level

left eyebrow, 50 % of the landmarks in the right eye and 37.5 % of the landmarks in the left eye respectively show a significant time effect. Percentage of significant landmarks in the mouth region starts to decrease when comparing times 3 and 4 (only 25 % of the landmarks in the mouth region shows a significant time effect). Most of the changes in the eye region occurs when comparing time frames 1 and 2, to a lesser extent, times 2 and 3. Hence in general we may conclude that surprise emotion and resulting facial expression is sudden and does not last long, since most of the changes in facial expression occurs immediately, within the first 2 time frames.

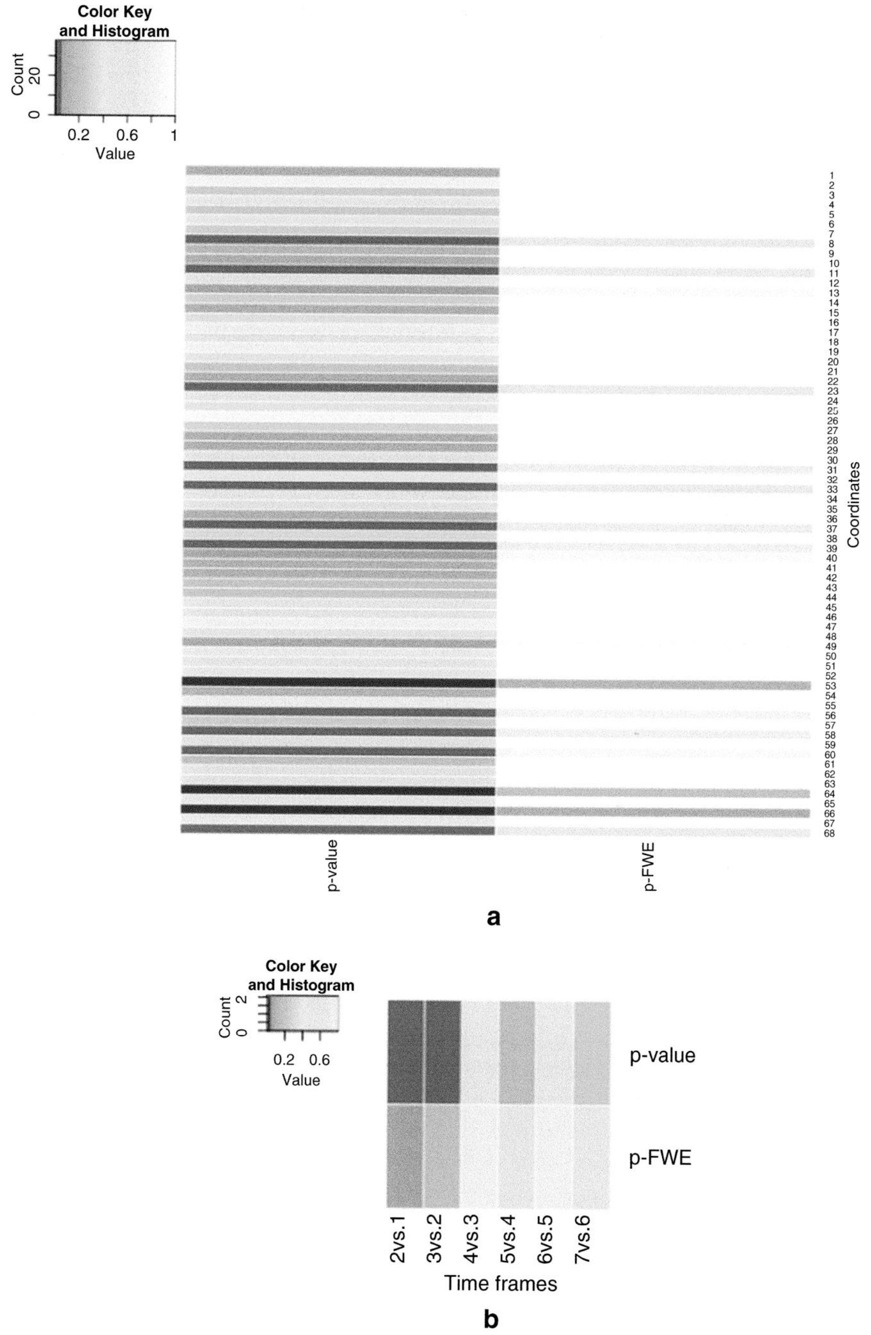

Fig. 5.16 Heat map representation of global *p*-values, obtained combining times (**a**) or coordinates (**b**)

Analyzing the heat map in Fig. 5.15, and considering all contrasts, we note that significant changes occur in the j_2 coordinates in mouth region (mouth opening), while in the right eyebrow and eye regions landmark coordinates are characterized by significant horizontal changes.

Moreover, the analysis suggests that changes along the j_2 coordinates occur especially when considering contrasts of time frames 3–4 and 4–5, and changes in the j_1 coordinates are more frequent at the beginning (time-frames 1–2 and 2–3) and at the end of the process (time-frames 6–7).

Once again, our results seems to confirm those found in Sect. 3.4.1, where vertical mouth changes were found to prevail in the generation of surprised facial expression. It must be noticed that the information provided by NPC test is more specific and detailed at each time point, at each contrast, hence our evaluations are referred to a discretized continuum process rather than to a continuum process, as done with polynomial shape modeling.

It must be emphasized that here we have focused on changes between consecutive times; however, different design matrices, which allow for all possible comparisons or evaluation of contrasts against the baseline, may be specified.

5.4 Introduction on Paired Landmark Data

Paired data issues in shape analysis context are often related to the study of symmetric structures. The most important type of symmetry in the organisation of living organisms is bilateral symmetry. A 2D (or 3D) object is said to be bilaterally symmetric if its mirror image about some line or some plane is the same as the original form after relabelling some landmarks. This mirroring locus in general is called the *midplane*. In a perfect bilaterally symmetric shape it is possible to distinguish two types of landmarks: *paired landmarks*, that do not lie on the midplane, but appear separately on left and right sides, and *unpaired landmarks*, that lie on the midplane. In the analysis of bilaterally symmetric structures, it is possible to identify two main types of symmetry: *matching symmetry* and *object symmetry*. Object symmetry relates to the symmetry within a single object, such as a human face, hence it considers parts with internal left-right symmetry. Matching symmetry has been introduced in Chap. 3 (Sect. 3.5). As anticipated, in matching symmetry two separate structures exist as mirror images of each other, one on each body side, e.g., left and right hand (Klingenberg et al. 2002; Mardia et al. 2000).

In order to study matching symmetry, the landmark configurations from one side are reflected, then all the configurations are superimposed by GPA to produce an overall mean shape. Variations in the averages of the pairs of configurations embody the symmetric variation among individuals. The deviations of each configuration from the consensus provide an estimate of the asymmetry component.

For the analysis of object symmetry, the data set includes both the original landmark configurations and their reflected copies with the paired landmarks relabelled. A GPA is applied to all configurations to produce a single consensus, which is symmetric. The symmetric variation among individuals is measured from the averages of the original configuration and its reflected (appropriately relabelled) copy. Again the asymmetry is estimated by the deviations of each configuration from the consensus (Savriama and Klingenberg 2006).

To evaluate object symmetry in the isotropic case, usually Procrustes ANOVA and Goodall's F tests are used. This method allows to identify and quantify different sources of shape variation: variation among individuals and sides (the so-called *directional asymmetry*), and variation due to an individual by side interaction (namely *fluctuating symmetry*).

These asymmetries convey interesting information on the evolutionary history, suggesting how symmetry is broken during development (Palmer 1996).

Directional asymmetry (also known as *fixed asymmetry*) emerges either when one side is larger than the other on average, or the larger member of a bilateral pair tends to be on the same side.

Fluctuating asymmetry is considered the most familiar of these asymmetries, providing a surprisingly convenient measure of developmental precision: the more precisely each side develops the greater the symmetry (Palmer and Strobeck 1997).

It bears information on environmental quality, stress, health or fitness. In the non-isotropic case, we may use T^2 Hotelling's test and the approximation to Fisher's F distribution.

The same holds for matching symmetry. Obviously there is a difference in the degrees of freedom of the tests. In the isotropic case we may preform ANOVA test, while in the non-isotropic case we carry out Hotelling's T^2. As usual, when the number of shape variables is greater than the most practical sample size, no formal T^2 can be computed and working under a permutation framework is recommended. In particular it is possible to use a permutation test for which the pivotal role of the Procrustes distance is retained but the distributional assumptions underlying the F under H_0 are relaxed. The reference distribution becomes a Monte Carlo permutation distribution where what is permuted is the assignment of one of the forms to the reflected state (Mardia et al. 2000).

In the nonparametric permutation framework, usually testing for symmetry corresponds to test the null hypothesis that a certain q-dimensional variable $\mathbf{V}$ is symmetric around 0, thus leading to solving problems for multivariate paired data observations (Pesarin 1990; Pesarin and Salmaso 2010).

The same framework applied in Sect. 5.3 to evaluate differences between facial expression at each time frame may be used to analyze object symmetry. Actually, we firstly consider all the differences between right and left coordinates of each landmark point. Then, once obtained partial p-values for the coordinates (coordinate level), we combine these p-values in order to obtain information on landmarks (landmark level). Finally we consider domains and aspects, if present, as well as the global combination of partial p-values.

5.5 Evaluating Symmetry Within Happy Facial Expression: Object Symmetry

In order to evaluate object symmetry, we applied the same superimposition procedure used to compare happy and surprised facial expression, replacing surprised data with reflected happy data and including a relabelling step. Even in this case, the neutral expression, i.e., data available on the first time frame for each subject, was used as the reference configuration to estimate the pole of the tangent projections. Procrustes residuals from the pole, which are approximate tangent coordinates to shape space, allow to represent, for each expression (happy and "reflected" happy), departures of each data shape from the neutral state. Then NPC methodology was applied to test symmetry for the following hypotheses

$$H_{0j_st}:\left\{V_{hj_s}(t)\stackrel{d}{=}V_{hrefj_s}(t)\right\},\ j_s=1,\ldots,q,\ t=1,2,\ldots,T,$$

then suitably combined to test the null hypothesis:

$$H_{0t}:\left\{\bigcap_{j_s=1}^{q}H_{0j_st}\right\},t=1,2,\ldots,T$$

$$H_{0j_s}:\left\{\bigcap_{t=1}^{T}H_{0j_st}\right\},j_s=1,2,\ldots,q$$

against the partial alternatives

$$H_{1j_st}:\left\{V_{hj_s}(t)\stackrel{d}{<\neq>}V_{hrefj_s}(t)\right\},$$

suitably combined in a global test, in which at least one null hypothesis is not true

$$H_{1t}:\left\{\bigcup_{j_s=1}^{q}H_{1j_st}\right\},\ t=1,2,\ldots,T$$

$$H_{1j_s}:\left\{\bigcup_{t=1}^{T}H_{1j_st}\right\},\ j_s=1,2,\ldots,q.$$

where V_h and V_{href} represent respectively data on happy and "reflected" happy facial expression and the test statistics of interest is obtained as seen in Sect. 5.3 and is defined as

$$T^*_{j_s}(t)=\sum_{i=1}^{N}\left[\bar{\mathbf{V}}_{h_i}(t)-\bar{\mathbf{V}}_{hrefi}(t)\right]\cdot\mathcal{S}^*_i,\ i=1,\ldots,N,\ t=1,2,\ldots,T;$$

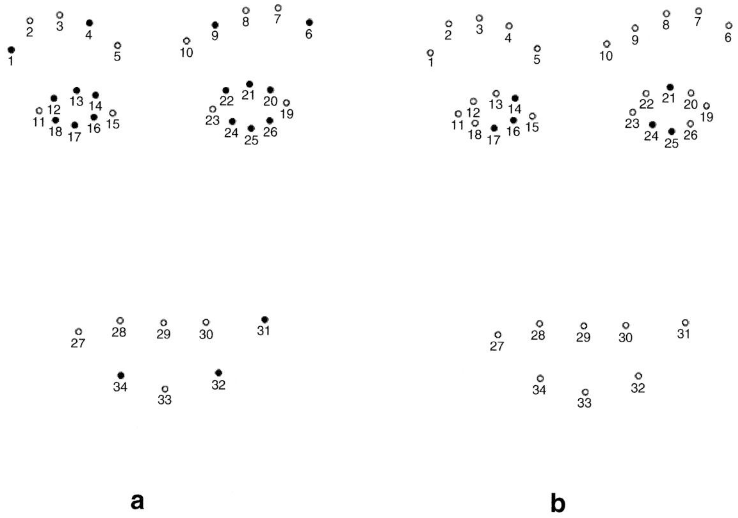

Fig. 5.17 Evaluating symmetry at landmark level. Overall effect is shown, after combining throughout times, before **(a)** and after **(b)** after FWE correction (points marked with a solid circle refer to landmarks with a significant *p*-value)

Results highlights that asymmetry prevails in the eye region (see Fig. 5.17) and remains significant even after adjustment for multiplicity. Globally happy facial expression is asymmetric.

Going into details, asymmetry in the eye regions is considerable for all the time frames, with the exception of time-frame 3. Mouth asymmetry is found only at time-frame 1 and time-frame 7 (the beginning and the end of the dynamic), with only 3 out 8 landmarks in the mouth region being significant before correction for multiplicity (see Figs. 5.18 and 5.19).

These results are in agreement with those presented in Sect. 3.5 suggesting that in general the mouth region in the happy facial expression is symmetric in its dynamics.

Examining results at coordinate levels (see Figs. 5.20 and 5.21), we may conclude that asymmetry mostly affects j_1 (horizontal) coordinates.

The key advantage of applying NPC methodology in assessing asymmetry is that we are able to identify asymmetric part of the face, thus giving a different (since it is given at landmark or coordinate level) and more precise evaluation of asymmetry that provided by global tests in ANOVA analysis.

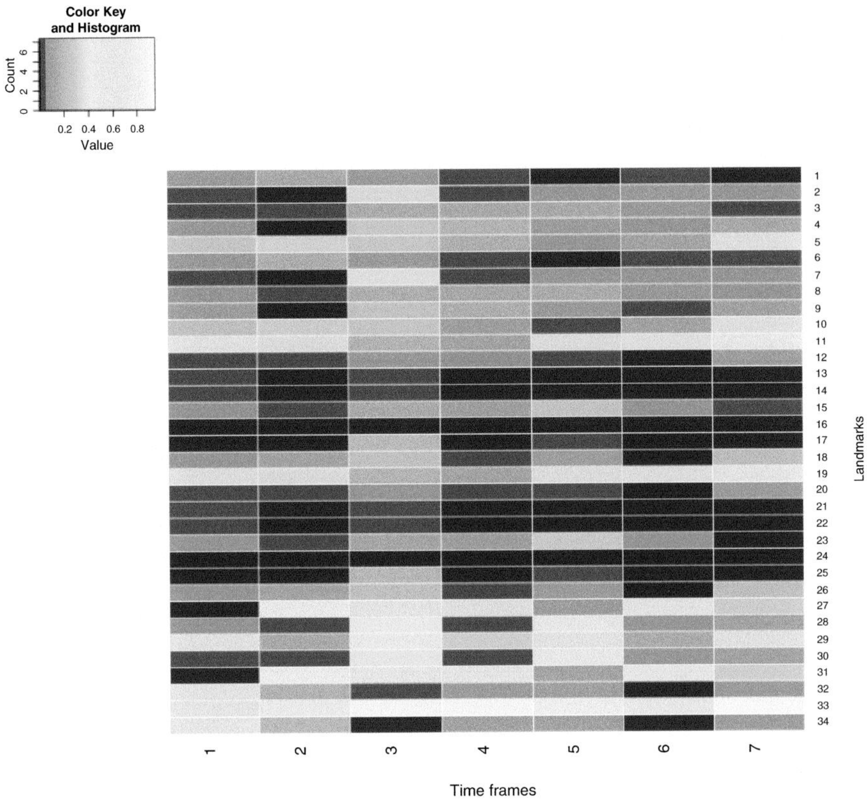

Fig. 5.18 Heat map representation of p-values. Evaluating symmetry within happy facial expressions at landmark level

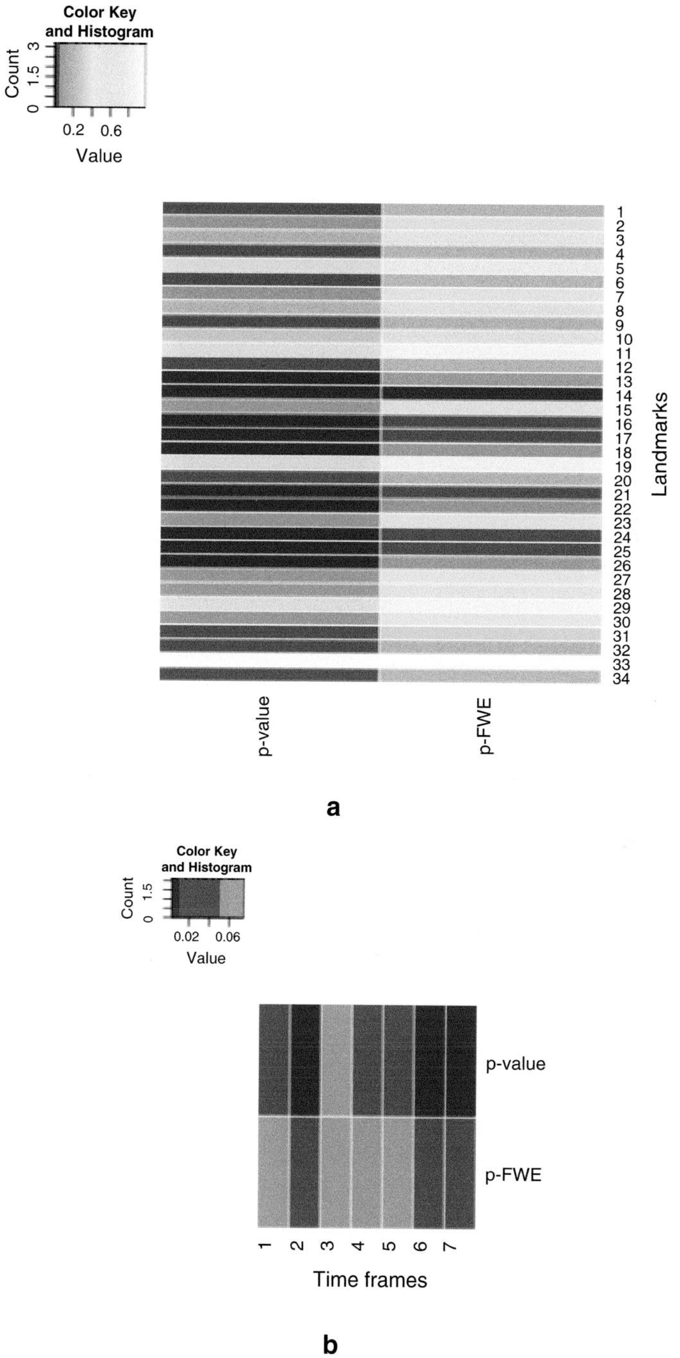

Fig. 5.19 Heat map representation of global p-values, obtained combining times (**a**) or landmarks (**b**)

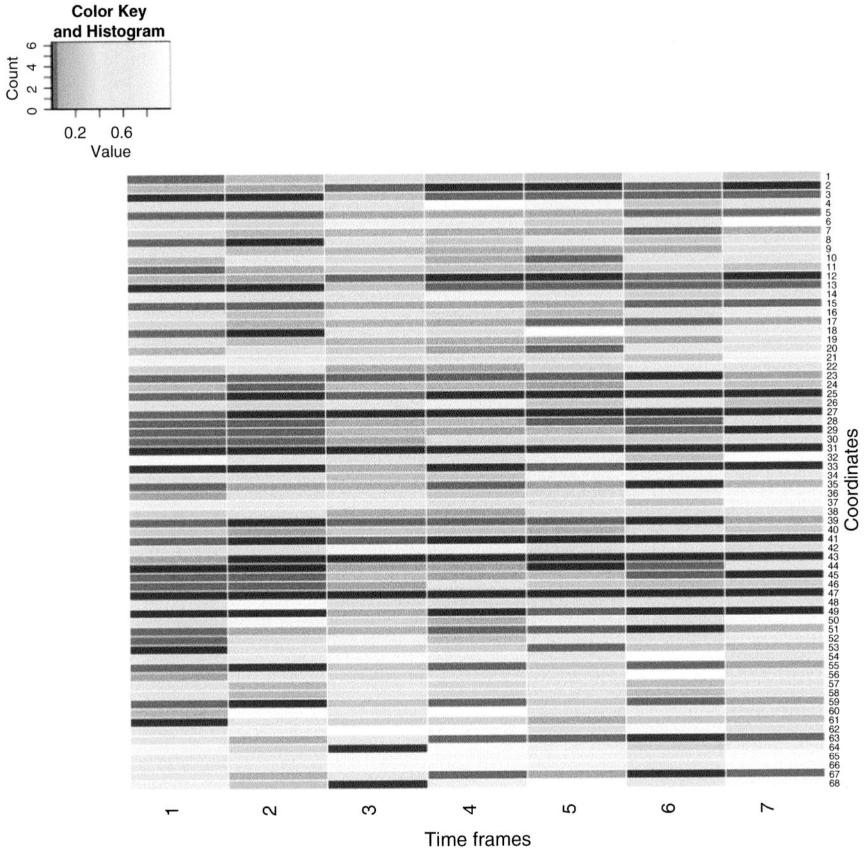

Fig. 5.20 Heat map representation of p-values. Evaluating symmetry within happy facial expressions at coordinates level

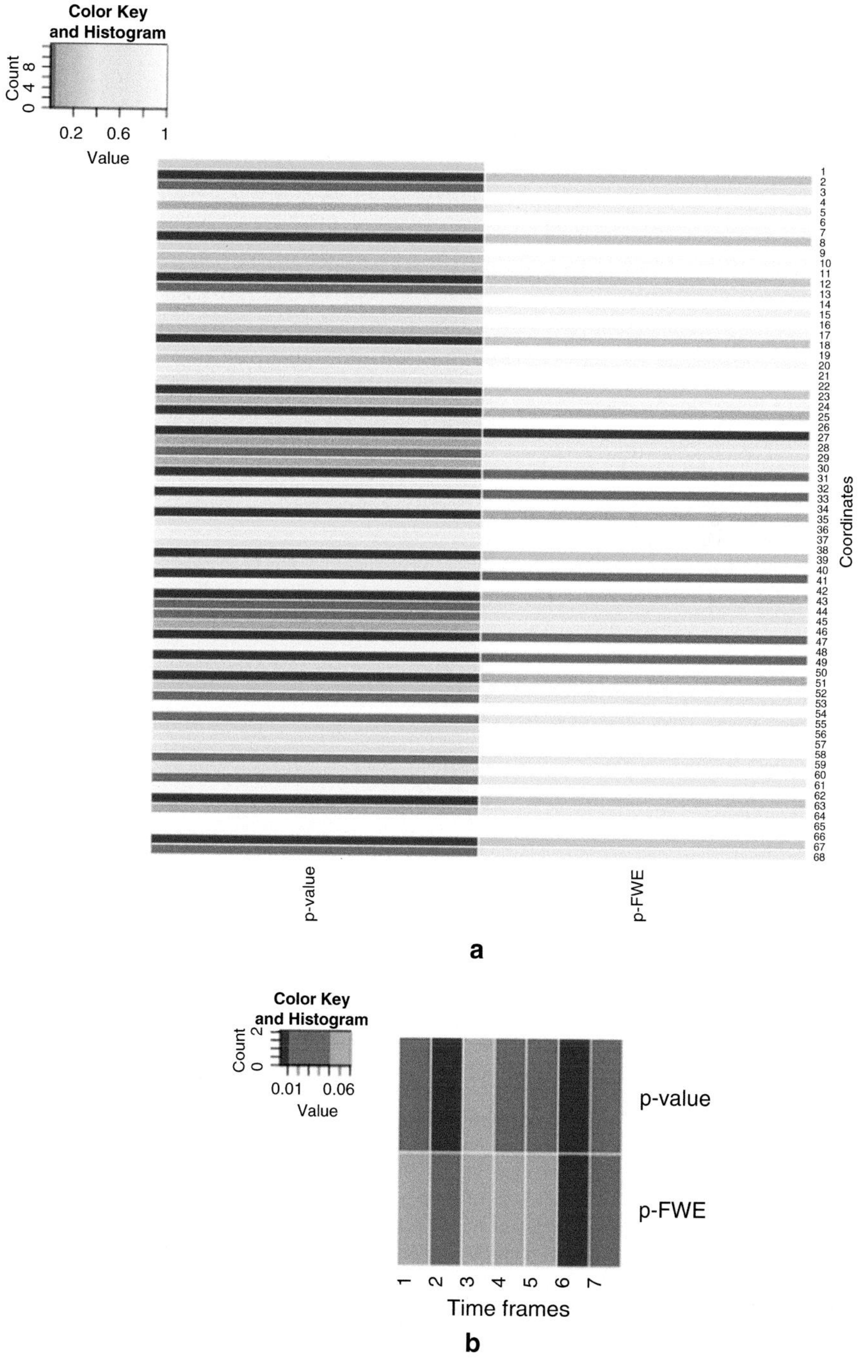

Fig. 5.21 Heat map representation of global *p*-values, obtained combining times (**a**) or coordinates (**b**)

References

Barry SJ (2008) Longitudinal analysis of three-dimensional facial shape data. Ph.D. thesis. University of Glasgow

Barry SJE, Bowman AW (2008) Linear mixed models for longitudinal shape data with applications to facial modeling. Biostatistics 9:555–565

Durrleman S, Pennec X, Trouvé A, Braga J, Gerig G, Ayache N (2013) Toward a comprehensive framework for the spatiotemporal statistical analysis of longitudinal shape data. Int J Comput Vis 103:22–59

Faraway J (1997) Regression analysis for a functional response. Technometrics 39:254–261

Fieuws S, Verbeke G (2006) Pairwise fitting of mixed models for the joint modelling of multivariate longitudinal profiles. Biometrics 62:424–431

Fishbaugh J, Prastawa M, Durrleman S, Piven J, Gerig G (2012) Analysis of longitudinal shape variability via subject specific growth modeling. In: Ayache N, Delingette H, Golland P, Mori K (eds) Medical image computing and computer-assisted intervention – MICCAI 2012. Lecture notes in computer science, vol 7510. Springer, Berlin, Heidelberg, pp 731–738

Fitzmaurice G, Laird N, Ware J (2011) Applied longitudinal analysis. Wiley, New York

Fontanella L, Ippoliti L, Valentini P (2013) A functional spatio-temporal model for geometric shape analysis. In: Torelli N, Pesarin F, Bar-Hen A (eds) Advances in theoretical and applied statistics. Springer, Berlin, pp 75–86

Kent J, Mardia K, Morris R, Aykroyd R (2001) Functional models of growth for landmark data. In: Mardia K, Kent J, Aykroyd R (eds) Proceedings in functional and spatial data analysis. Springer, Berlin, pp 75–86

Klingenberg CP, Barluenga M, Meyer A (2002) Shape analysis of symmetric structures: quantifying variation among individuals and asymmetry. Evolution 56:1909–1920

Kume A (2000) Shape space smoothing splines for planar landmark data. Biometrika 94:513–528

Laird NM, Ware J (1982) Random-effects models for longitudinal data. Biometrics 38:963–974

Le H, Kume A (2000) Detection of shape changes in biological features. J Microsc 2:140–147

Mardia KV, Bookstein FL, Moreton IJ (2000) Statistical assessment of bilateral symmetry of shapes. Biometrika 87:285–300

Palmer AR (1996) From symmetry to asymmetry: phylogenetic patterns of asymmetry variation in animals and their evolutionary significance. Proc Natl Acad Sci 93:14,279–14,286

Palmer AR, Strobeck C (1997) Fluctuating asymmetry and developmental stability: heritability of observable variation vs. heritability of inferred cause. J Evol Biol 10:39–49

Pantic M, Cowie R, Drrico F, Heylen D, Mehu M, Pelachaud P (2011) Social signal processing: the research agenda. In: Visual analysis of humans. Springer, London, pp 511–538

Pesarin F (1990) On a nonparametric combination method for dependent permutation tests with applications. Psychother Psychosom 54:172–179

Pesarin F, Salmaso L (2010) Permutation tests for complex data: theory, applications and software. Wiley, New York

Pinheiro J, Bates D (2000) Mixed-effects models in S and S-PLUS. Springer, Berlin

Savriama Y, Klingenberg CP (2006) Geometric morphometrics of complex symmetric structures: Shape analysis of symmetry and asymmetry with procrustes methods. In: Barber S, Baxter PD, Mardia KV, Walls RE (eds) Interdisciplinary Statistics and bioinformatics. Leeds University Press, Leeds, pp 158–161. ISBN 0-85316-252-2

Verbeke G, Fieuws S (2005) Evaluation of the pairwise approach for fitting joint linear mixed models: a simulation study. Technical Report TR0527, Biostatistical Centre, Katholieke Universiteit Leuven, Belgium

Verbeke G, Molenberghs G (2000) Linear mixed models for longitudinal data. Springer, New York

Verbeke G, Molenberghs G, Rizopoulos D (2010) Random effects models for longitudinal data. In: van Montfort K, Oud JHL, Satorra A (eds) Longitudinal research with latent variables. Springer, Berlin, Heidelberg, pp 37–96

Vinciarelli A, Pantic M, Bourlard H (2009) Social signal processing: survey of an emerging domain. Image Vis Comput 27:1743–1759

Appendix A
Shape Inference and the Offset-Normal Distribution

A.1 EM Algorithm for Estimating μ and Σ

The conditional distribution of **X** given its shape **u**, by applying Bayes theorem, can be written as

$$dF\left(\mathbf{X}|\mathbf{u},\boldsymbol{\mu},\boldsymbol{\Sigma}\right)=\frac{f\left(\mathbf{h},\mathbf{u};\boldsymbol{\mu},\boldsymbol{\Sigma}\right)d\mathbf{h}}{f\left(\mathbf{u};\boldsymbol{\mu},\boldsymbol{\Sigma}\right)}$$

Given the transformation $vec\left(\mathbf{X}\right)=\mathbf{W}\mathbf{h}$ and considering the change of variable, $\mathbf{h}=\boldsymbol{\Psi}\mathbf{l}$, such that $\mathbf{l}\sim\mathcal{N}(\boldsymbol{\zeta},\mathbf{D})$ with $\mathbf{D}=diag(\sigma_1,\sigma_2)$, we have:

$$\frac{\int\mathbf{W}\mathbf{h}f\left(\mathbf{h},\mathbf{u};\boldsymbol{\mu},\boldsymbol{\Sigma}\right)d\mathbf{h}}{f\left(\mathbf{u};\boldsymbol{\mu},\boldsymbol{\Sigma}\right)}=\mathbf{W}\boldsymbol{\Psi}\frac{\int\mathbf{l}f(\mathbf{l},\mathbf{u};\boldsymbol{\mu},\boldsymbol{\Sigma})d\mathbf{l}}{f(\mathbf{u};\boldsymbol{\mu},\boldsymbol{\Sigma})}$$

$$\frac{\int\mathbf{W}\mathbf{h}\mathbf{h}'\mathbf{W}'f\left(\mathbf{h},\mathbf{u};\boldsymbol{\mu},\boldsymbol{\Sigma}\right)d\mathbf{h}}{f\left(\mathbf{u};\boldsymbol{\mu},\boldsymbol{\Sigma}\right)}=\mathbf{W}\boldsymbol{\Psi}\frac{\int\mathbf{l}\mathbf{l}'f(\mathbf{l},\mathbf{u};\boldsymbol{\mu},\boldsymbol{\Sigma})d\mathbf{l}}{f(\mathbf{u};\boldsymbol{\mu},\boldsymbol{\Sigma})}\boldsymbol{\Psi}'\mathbf{W}'.$$

As in Eq. (2.5) we thus have

$$f\left(\mathbf{u};\boldsymbol{\mu},\boldsymbol{\Sigma}\right)=M\sum_{j=0}^{K-2}\binom{K-2}{j}E[l_1^{2j}|\zeta_1,\sigma_1]E[l_2^{2(K-2-j)}|\zeta_2,\sigma_2]$$

where $M=\frac{|\boldsymbol{\Gamma}|^{\frac{1}{2}}exp(-g/2)}{(2\pi)^{K-2}|\boldsymbol{\Sigma}|^{\frac{1}{2}}}$.

C. Brombin et al., *Parametric and Nonparametric Inference for Statistical Dynamic Shape Analysis with Applications*, SpringerBriefs in Statistics,
DOI 10.1007/978-3-319-26311-3

For the update of the elements of the mean vector, we have to consider the following integrals

$$\int l_1 f(\mathbf{l}, \mathbf{u}; \boldsymbol{\mu}, \boldsymbol{\Sigma}) d\mathbf{l} = M \sum_{j=0}^{K-2} \binom{K-2}{j} \int l_1^{2j+1} f_{\mathcal{N}}(l_1; \zeta_1, \sigma_1) dl_1 \int l_2^{2(K-2-j)} f_{\mathcal{N}}(l_2; \zeta_2, \sigma_2) dl_2 =$$

$$= M \sum_{j=0}^{K-2} \binom{K-2}{j} E[l_1^{2j+1} | \zeta_1, \sigma_1] E[l_2^{2(K-2-j)} | \zeta_2, \sigma_2]$$

$$\int l_2 f(\mathbf{l}, \mathbf{u}; \boldsymbol{\mu}, \boldsymbol{\Sigma}) d\mathbf{l} = M \sum_{j=0}^{K-2} \binom{K-2}{j} \int l_1^{2j} f_{\mathcal{N}}(l_1; \zeta_1, \sigma_1) dl_1 \int l_2^{2(K-2-j)+1} f_{\mathcal{N}}(l_2; \zeta_2, \sigma_2) dl_2 =$$

$$= M \sum_{j=0}^{K-2} \binom{K-2}{j} E[l_1^{2j} | \zeta_1, \sigma_1] E[l_2^{2(K-2-j)+1} | \zeta_2, \sigma_2].$$

Analogously, for the update rules of the covariance elements we have

$$\int_{\mathbb{R}^2} l_1^2 f(\mathbf{l}, \mathbf{u}; \boldsymbol{\mu}, \boldsymbol{\Sigma}) d\mathbf{l} = M \sum_{j=0}^{K-2} \binom{K-2}{j} \int l_1^{2j+2} f_{\mathcal{N}}(l_1; \zeta_1, \sigma_1) dl_1 \int l_2^{2(K-2-j)} f_{\mathcal{N}}(l_2; \zeta_2, \sigma_2) dl_2 =$$

$$= M \sum_{j=0}^{K-2} \binom{K-2}{j} E[l_1^{2j+2} | \zeta_1, \sigma_1] E[l_2^{2(K-2-j)} | \zeta_2, \sigma_2]$$

$$\int l_2^2 f(\mathbf{l}, \mathbf{u}; \boldsymbol{\mu}, \boldsymbol{\Sigma}) d\mathbf{l} = M \sum_{j=0}^{K-2} \binom{K-2}{j} \int l_1^{2j} f_{\mathcal{N}}(l_1; \zeta_1, \sigma_1) dl_1 \int l_2^{2(K-2-j)+2} f_{\mathcal{N}}(l_2; \zeta_2, \sigma_2) dl_2$$

$$= M \sum_{j=0}^{K-2} \binom{K-2}{j} E[l_1^{2j} | \zeta_1, \sigma_1] E[l_2^{2(K-2-j)+2} | \zeta_2, \sigma_2]$$

$$\int l_1 l_2 f(\mathbf{l}, \mathbf{u}; \boldsymbol{\mu}, \boldsymbol{\Sigma}) d\mathbf{l} = M \sum_{j=0}^{K-2} \binom{K-2}{j} \int l_1^{2j+1} f_{\mathcal{N}}(l_1; \zeta_1, \sigma_1) dl_1 \int l_2^{2(K-2-j)+1} f_{\mathcal{N}}(l_2; \zeta_2, \sigma_2) dl_2$$

$$= M \sum_{j=0}^{K-2} \binom{K-2}{j} E[l_1^{2j+1} | \zeta_1, \sigma_1] E[l_2^{2(K-2-j)+1} | \zeta_2, \sigma_2].$$

Notice that the term $M = \frac{|\boldsymbol{\Gamma}|^{1/2} \exp\{-g/2\}}{(2\pi)^{K-2} |\boldsymbol{\Sigma}|^{1/2}}$ is present in $f(\mathbf{u}; \boldsymbol{\mu}, \boldsymbol{\Sigma})$, $\int \mathbf{l} f(\mathbf{l}, \mathbf{u}; \boldsymbol{\mu}, \boldsymbol{\Sigma}) d\mathbf{l}$ and $\int \mathbf{l}\mathbf{l}' f(\mathbf{l}, \mathbf{u}; \boldsymbol{\mu}, \boldsymbol{\Sigma}) d\mathbf{l}$, but it cancels out in the ratios (2.11) and so the required expressions are obtained in terms of univariate Gaussian expectations—see, Eq. (2.6).

A.2 EM for Complex Covariance

Given the complex number $z_2 = x_2 + iy_2$, the rotation and scale parameter vector, $\mathbf{h} = (\mathscr{R}e(z_2)\ \mathscr{I}m(z_2)) = (x_2\ y_2)'$, has covariance matrix $\boldsymbol{\Gamma} = (\mathbf{W}'\boldsymbol{\Sigma}^{-1}\mathbf{W})^{-1}$ (see Sect. 2.4). It can be shown that if $\boldsymbol{\Sigma}$ has a complex structure, $\boldsymbol{\Gamma}$ is a multiple of the identity matrix, i.e. $\boldsymbol{\Gamma} = \frac{\gamma_z}{2}\mathbf{I}_2$. Let $\mathbf{W}$ be written as

$$\mathbf{W} = \begin{pmatrix} \mathbf{u} & -\mathbf{v} \\ \mathbf{v} & \mathbf{u} \end{pmatrix}$$

and consider that the precision matrix $\boldsymbol{Q} = \boldsymbol{\Sigma}^{-1}$ is given by

$$\boldsymbol{\Sigma}^{-1} = \begin{pmatrix} \mathbf{Q}_1 & \mathbf{Q}_2 \\ -\mathbf{Q}_2 & \mathbf{Q}_1 \end{pmatrix}$$

where the symmetric matrix $\mathbf{Q}_1 = (\mathbf{C}_1 + \mathbf{C}_2\mathbf{C}_1^{-1}\mathbf{C}_2)^{-1}$ and the skew-symmetric matrix $\mathbf{Q}_2 = \mathbf{Q}_1\mathbf{C}_2\mathbf{C}_1^{-1}$ are respectively the real and imaginary parts of the inverse of the complex covariance in the pre-form space: $\mathbf{Q}_z = \boldsymbol{\Sigma}_z^{-1} = \frac{1}{2}(\mathbf{Q}_1 + i\mathbf{Q}_2)$. Therefore

$$\begin{aligned}
\boldsymbol{\Gamma}^{-1} = \mathbf{W}'\boldsymbol{\Sigma}^{-1}\mathbf{W} &= \begin{pmatrix} \mathbf{u}' & \mathbf{v}' \\ -\mathbf{v}' & \mathbf{u}' \end{pmatrix} \begin{pmatrix} \mathbf{Q}_1 & \mathbf{Q}_2 \\ -\mathbf{Q}_2 & \mathbf{Q}_1 \end{pmatrix} \begin{pmatrix} \mathbf{u} & -\mathbf{v} \\ \mathbf{v} & \mathbf{u} \end{pmatrix} = \\
&= \begin{pmatrix} \mathbf{u}'\mathbf{Q}_1\mathbf{u} - \mathbf{v}'\mathbf{Q}_2\mathbf{u} + \mathbf{u}'\mathbf{Q}_2\mathbf{v} + \mathbf{v}'\mathbf{Q}_1\mathbf{v} & -\mathbf{u}'\mathbf{Q}_1\mathbf{v} + \mathbf{v}'\mathbf{Q}_2\mathbf{v} + \mathbf{u}'\mathbf{Q}_2\mathbf{u} + \mathbf{v}'\mathbf{Q}_1\mathbf{u} \\ -\mathbf{v}'\mathbf{Q}_1\mathbf{u} + \mathbf{u}'\mathbf{Q}_2\mathbf{u} - \mathbf{v}'\mathbf{Q}_2\mathbf{v} + \mathbf{u}'\mathbf{Q}_1\mathbf{v} & \mathbf{v}'\mathbf{Q}_1\mathbf{v} + \mathbf{u}'\mathbf{Q}_2\mathbf{v} - \mathbf{v}'\mathbf{Q}_2\mathbf{u} + \mathbf{u}'\mathbf{Q}_1\mathbf{u} \end{pmatrix} \\
&= \begin{pmatrix} \mathbf{u}'\mathbf{Q}_1\mathbf{u} + \mathbf{v}'\mathbf{Q}_1\mathbf{v} + 2\mathbf{u}'\mathbf{Q}_2\mathbf{v} & 0 \\ 0 & \mathbf{u}'\mathbf{Q}_1\mathbf{u} + \mathbf{v}'\mathbf{Q}_1\mathbf{v} + 2\mathbf{u}'\mathbf{Q}_2\mathbf{v} \end{pmatrix}
\end{aligned}$$

In the solution we have exploited the properties of symmetric and skew-symmetric matrices: $\mathbf{u}'\mathbf{Q}_1\mathbf{v} = \mathbf{v}'\mathbf{Q}_1\mathbf{u}$, $\mathbf{u}'\mathbf{Q}_2\mathbf{u} = \mathbf{v}'\mathbf{Q}_2\mathbf{v} = 0$ and $\mathbf{u}'\mathbf{Q}_2\mathbf{v} = -\mathbf{v}'\mathbf{Q}_2\mathbf{u}$.

It's easy to show that

$$\gamma_z^{-1} = \boldsymbol{\xi}^*\boldsymbol{\Sigma}_z^{-1}\boldsymbol{\xi} = \frac{1}{2}(\mathbf{u} + i\mathbf{v})^*(\mathbf{Q}_1 + i\mathbf{Q}_2)(\mathbf{u} + i\mathbf{v}) = \frac{1}{2}(\mathbf{u}'\mathbf{Q}_1\mathbf{u} + \mathbf{v}'\mathbf{Q}_1\mathbf{v} + 2\mathbf{u}'\mathbf{Q}_2\mathbf{v})$$

and, therefore, $\boldsymbol{\Gamma}^{-1} = 2\gamma_z^{-1}\mathbf{I}_2$ and $\boldsymbol{\Gamma} = \frac{\gamma_z}{2}\mathbf{I}_2$.

As a result writing $z_2 = h_1 + ih_2$, $\eta = \eta_1 + i\eta_2$, $\sigma^2 = \gamma_z/2$, and considering that h_1 and h_2 are independent, the real part of

$$\int z_2\, \|z_2\|^{2(K-2)} f_{\mathscr{CN}}\left(z_2, \boldsymbol{\xi}; \boldsymbol{\mu}_z, \boldsymbol{\Sigma}_z\right) dz_2$$

is now

$$\int h_1(h_1^2+h_2^2)^{K-2} f_{\mathcal{N}}(\mathbf{h},\mathbf{u};\boldsymbol{\mu},\boldsymbol{\Sigma})\,d\mathbf{h}$$

$$= \frac{\sigma^2 e^{-g_z}}{\pi^{(K-2)}|\boldsymbol{\Sigma}_z|}\sum_{k=0}^{K-2}\binom{K-2}{k}\int h_1 h_1^{2k} f_{\mathcal{N}}(h_1;\eta_1,\sigma) h_2^{2(K-2-k)} f_{\mathcal{N}}(h_2;\eta_2,\sigma)d\mathbf{h} =$$

$$= \frac{\sigma^2 e^{-g_z}}{\pi^{(K-2)}|\boldsymbol{\Sigma}_z|}\sum_{k=0}^{K-2}\binom{K-2}{k} E[h_1^{2k+1}|\eta_1,\sigma]E[h_2^{2(K-2-k)}|\eta_2,\sigma]$$

Applying the addition formula given in 8.974/4 in Gradshteyn and Ryzhik (1980)

$$\sum_{j=0}^{m}\mathcal{L}_j^{(\alpha)}(x)\mathcal{L}_{m-j}^{(\beta)}(y) = \mathcal{L}_m^{(\alpha+\beta+1)}(x+y)$$

and the relation between Laguerre polynomials and Hermite polynomials for Gaussian moments, we have

$$\sum_{k=0}^{K-2}\binom{K-2}{k} E[h_1^{2k+1}|\eta_1,\sigma]Eh_2^{2(K-2-k)}|\eta_2,\sigma]$$

$$= \sum_{k=0}^{K-2}\binom{K-2}{k}\left[\eta_1(2\sigma^2)^k k!\mathcal{L}_k^{(1/2)}\left(-\frac{\eta_1^2}{2\sigma^2}\right)\right]\left[(2\sigma^2)^{K-2-k}(K-2-k)!\mathcal{L}_{K-2-k}^{(-1/2)}\left(-\frac{\eta_2^2}{2\sigma^2}\right)\right]$$

$$= (K-2)!(2\sigma^2)^{K-2}\eta_1\sum_{k=0}^{K-2}\mathcal{L}_k^{(1/2)}\left(-\frac{\eta_1^2}{2\sigma^2}\right)\mathcal{L}_{K-2-k}^{(-1/2)}\left(-\frac{\eta_2^2}{2\sigma^2}\right)$$

$$= (K-2)!\gamma_z^{K-2}\eta_1\mathcal{L}_{K-2}^{(1)}\left(-\frac{\|\eta\|^2}{\gamma_z}\right).$$

Analogously, the imaginary part of $\int z_2\,\|z_2\|^{2(K-2)} f_{\mathcal{CN}}\left(z_2,\boldsymbol{\xi};\boldsymbol{\mu}_z,\boldsymbol{\Sigma}_z\right)dz_2$ is

$$\int h_2\,\|\mathbf{h}\|^{2(K-2)} f_{\mathcal{N}}(\mathbf{h},\mathbf{u};\boldsymbol{\mu},\boldsymbol{\Sigma})d\mathbf{h}$$

$$= \frac{\sigma^2 e^{-g_z}}{\pi^{(K-2)}|\boldsymbol{\Sigma}_z|}\sum_{k=0}^{K-2}\binom{K-2}{k}\int h_1^{2k} f_{\mathcal{N}}(h_1;\eta_1,\sigma) h_2 h_2^{2(K-2-k)} f_{\mathcal{N}}(h_2;\eta_2,\sigma)d\mathbf{h}$$

$$= \frac{\sigma^2 e^{-g_z}}{\pi^{(K-2)}|\boldsymbol{\Sigma}_z|}\sum_{k=0}^{K-2}\binom{K-2}{k} E[h_1^{2k}|\eta_1,\sigma]E[h_2^{2(K-2-k)+1}|\eta_2,\sigma]$$

where

$$\sum_{k=0}^{K-2}\binom{K-2}{k}E[h_1^{2k}|\eta_1,\sigma]E[h_2^{2(K-2-k)+1}|\eta_2,\sigma]$$

$$=\sum_{k=0}^{K-2}\binom{K-2}{k}(2\sigma^2)^k k!\mathscr{L}_k^{(-1/2)}\left(-\frac{\eta_1^2}{2\sigma^2}\right)\eta_2(2\sigma^2)^{K-2-k}(K-2-k)!\mathscr{L}_{K-2-k}^{(1/2)}\left(-\frac{\eta_2^2}{2\sigma^2}\right)$$

$$=(K-2)!(2\sigma^2)^{K-2}\eta_2\sum_{k=0}^{K-2}\mathscr{L}_k^{(-1/2)}\left(-\frac{\eta_1^2}{2\sigma^2}\right)\mathscr{L}_{K-2-k}^{(1/2)}\left(-\frac{\eta_2^2}{2\sigma^2}\right)$$

$$=(K-2)!\gamma_z^{K-2}\eta_2\mathscr{L}_{K-2}^{(1)}\left(-\frac{\|\eta\|^2}{\gamma_z}\right).$$

Recalling that $z_2 = h_1 + ih_2$ and $\eta = \eta_1 + i\eta_2$, we have

$$\int z_2\|z_2\|^{2(K-2)}f_{\mathscr{CN}}(z_2,\boldsymbol{\xi};\boldsymbol{\mu}_z,\boldsymbol{\Sigma}_z)dz_2$$

$$=\frac{\sigma^2e^{-g_z}}{\pi^{(K-2)}|\boldsymbol{\Sigma}_z|}\int h_1f_{\mathscr{N}}(h_1;\eta_1,\sigma)f_{\mathscr{N}}(h_2;\eta_2,\sigma)\|\mathbf{h}\|^{2(K-2)}d\mathbf{h}$$

$$+i\frac{\sigma^2e^{-g_z}}{\pi^{(K-2)}|\boldsymbol{\Sigma}_z|}\int h_2f_{\mathscr{N}}(h_1;\eta_1,\sigma)f_{\mathscr{N}}(h_2;\eta_2,\sigma)\|\mathbf{h}\|^{2(K-2)}d\mathbf{h}=$$

$$=\frac{\sigma^2e^{-g_z}}{\pi^{(K-2)}|\boldsymbol{\Sigma}_z|}(K-2)!\gamma_z^{K-2}\mathscr{L}_{K-2}^{(1)}\left(-\frac{\|\eta\|^2}{2\sigma^2}\right)(\eta_1+i\eta_2)$$

$$=\frac{\gamma_ze^{-g_z}}{2\pi^{(K-2)}|\boldsymbol{\Sigma}_z|}(K-2)!\eta\gamma_z^{K-2}\mathscr{L}_{K-2}^{(1)}\left(-\frac{\|\eta\|^2}{\gamma_z}\right).$$

From 8.971/4 of Gradshteyn and Ryzhik (1980) for which

$$\mathscr{L}_{K-2}^{(1)}(x)=\frac{K-1}{x}\left(\mathscr{L}_{K-2}(x)-\mathscr{L}_{K-1}(x)\right)$$

we have

$$\int z_2\|z_2\|^{2(K-2)}f_{\mathscr{CN}}(z_2,\boldsymbol{\xi};\boldsymbol{\mu}_z,\boldsymbol{\Sigma}_z)dz_2=$$

$$=\frac{\gamma_ze^{-g_z}}{2\pi^{(K-2)}|\boldsymbol{\Sigma}_z|}(K-2)!\eta\gamma_z^{K-2}(K-1)\left(-\frac{\gamma_z}{\|\eta\|^2}\right)\left[\mathscr{L}_{K-2}\left(-\frac{\|\eta\|^2}{\gamma_z}\right)-\mathscr{L}_{K-1}\left(-\frac{\|\eta\|^2}{\gamma_z}\right)\right]=$$

$$=\frac{\gamma_ze^{-g_z}}{2\pi^{(K-2)}|\boldsymbol{\Sigma}_z|}(K-1)!\omega\frac{\gamma_z^{K-2}}{\|\eta\|}\left[\mathscr{L}_{K-1}\left(-\frac{\|\eta\|^2}{\gamma_z}\right)-\mathscr{L}_{K-2}\left(-\frac{\|\eta\|^2}{\gamma_z}\right)\right]$$

where $\omega = \gamma_z\eta/\|\eta\|$.

Similarly, one can show that

$$\int \|z_2\|^{2(K-2)} f_{\mathscr{C}\mathscr{N}}(z_2, \boldsymbol{\xi}; \boldsymbol{\mu}_z, \boldsymbol{\Sigma}_z) dz_2$$

$$= \frac{\sigma^2 e^{-g_z}}{\pi^{(K-2)}|\boldsymbol{\Sigma}_z|} \sum_{k=0}^{K-2} \binom{K-2}{k} E[h_1^{2k}|\eta_1, \sigma] E[h_2^{2(K-2-k)}|\eta_2, \sigma]$$

$$= \frac{\sigma^2 e^{-g_z}}{\pi^{(K-2)}|\boldsymbol{\Sigma}_z|} \sum_{k=0}^{K-2} \binom{K-2}{k} \left[(2\sigma^2)^k k! \mathscr{L}_k^{(-1/2)}\left(-\frac{\eta_1^2}{2\sigma^2}\right)\right]$$

$$\times \left[(2\sigma^2)^{K-2-k}(K-2-k)! \mathscr{L}_{K-2-k}^{(-1/2)}\left(-\frac{\eta_2^2}{2\sigma^2}\right)\right]$$

$$= \frac{\sigma^2 e^{-g_z}}{\pi^{(K-2)}|\boldsymbol{\Sigma}_z|} \sum_{k=0}^{K-2} \frac{(K-2)!}{k!(K-2-k)!} k!(K-2-k)!$$

$$\times (2\sigma^2)^k (2\sigma^2)^{K-2-k} \mathscr{L}_k^{(-1/2)}\left(-\frac{\eta_1^2}{2\sigma^2}\right) \mathscr{L}_{K-2-k}^{(-1/2)}\left(-\frac{\eta_2^2}{2\sigma^2}\right)$$

$$= \frac{\sigma^2 e^{-g_z}}{\pi^{(K-2)}|\boldsymbol{\Sigma}_z|}(K-2)!(2\sigma^2)^{K-2} \mathscr{L}_{K-2}\left(-\frac{\|\eta\|^2}{2\sigma^2}\right)$$

$$= \frac{\gamma_z e^{-g_z}}{2\pi^{(K-2)}|\boldsymbol{\Sigma}_z|}(K-2)! \gamma_z^{K-2} \mathscr{L}_{K-2}\left(-\frac{\|\eta\|^2}{\gamma_z}\right)$$

and

$$\int \|z_2\|^2 \|z_2\|^{2(K-2)} f_{\mathscr{C}\mathscr{N}}(z_2, \boldsymbol{\xi}; \boldsymbol{\mu}_z, \boldsymbol{\Sigma}_z) dz_2 = \int \|z_2\|^{2(K-1)} f_{\mathscr{C}\mathscr{N}}(z_2, \boldsymbol{\xi}; \boldsymbol{\mu}_z, \boldsymbol{\Sigma}_z) dz_2$$

$$= \frac{\sigma^2 e^{-g_z}}{\pi^{(K-2)}|\boldsymbol{\Sigma}_z|} \sum_{k=0}^{K-1} \binom{K-1}{k} E[h_1^{2k}|\eta_1, \sigma] E[h_2^{2(K-1-k)}|\eta_2, \sigma]$$

$$= \frac{\sigma^2 e^{-g_z}}{\pi^{(K-2)}|\boldsymbol{\Sigma}_z|}(K-1)!(2\sigma^2)^{K-1} \mathscr{L}_{K-1}\left(-\frac{\|\eta\|^2}{2\sigma^2}\right)$$

$$= \frac{\gamma_z e^{-g_z}}{2\pi^{(K-1)}|\boldsymbol{\Sigma}_z|}(K-1)! \gamma_z^{K-1} \mathscr{L}_{K-1}\left(-\frac{\|\eta\|^2}{\gamma_z}\right).$$

Eqs. (2.15) and (2.16) follow from these expressions with the term $\frac{\gamma_z e^{-g_z}}{2\pi^{(K-1)}|\boldsymbol{\Sigma}_z|}$ canceling out. More specifically, for the update of the mean—Eq. (2.15)—we have

$$\frac{\int z_2 \|z_2\|^{2(K-2)} f_{\mathscr{C}\mathscr{N}}\left(z_2, \boldsymbol{\xi}^{(n)}; \boldsymbol{\mu}_z^{(r)}, \boldsymbol{\Sigma}_z^{(r)}\right) dz_2}{\int \|z_2\|^{2(K-2)} f_{\mathscr{C}\mathscr{N}}\left(z_2, \boldsymbol{\xi}^{(n)}; \boldsymbol{\mu}_z^{(r)}, \boldsymbol{\Sigma}_z^{(r)}\right) dz_2}$$

$$= \frac{\frac{\gamma_z e^{-g_z}}{2\pi^{(K-2)}|\boldsymbol{\Sigma}_z|}(K-1)!\omega\frac{\gamma_z^{K-2}}{\|\eta\|}\left[\mathscr{L}_{K-1}\left(-\frac{\|\eta\|^2}{\gamma_z}\right) - \mathscr{L}_{K-2}\left(-\frac{\|\eta\|^2}{\gamma_z}\right)\right]}{\frac{\gamma_z e^{-g_z}}{2\pi^{(K-2)}|\boldsymbol{\Sigma}_z|}(K-2)!\gamma_z^{K-2}\mathscr{L}_{K-2}\left(-\frac{\|\eta\|^2}{\gamma_z}\right)}$$

$$= \frac{\omega(K-1)}{\|\eta\|}\left(\frac{\mathscr{L}_{K-1}(\|\eta\|^2/\gamma_z)}{\mathscr{L}_{K-2}(\|\eta\|^2/\gamma_z)} - 1\right)$$

while for the update of the covariance—Eq. (2.16)—

$$\frac{\int \|z_2\|^2 \|z_2\|^{2(K-2)} f_{\mathscr{C}\mathscr{N}}\left(z_2, \boldsymbol{\xi}^{(n)}; \boldsymbol{\mu}_z^{(r)}, \boldsymbol{\Sigma}_z^{(r)}\right) dz_2}{\int \|z_2\|^{2(K-2)} f_{\mathscr{C}\mathscr{N}}\left(z_2, \boldsymbol{\xi}^{(n)}; \boldsymbol{\mu}_z^{(r)}, \boldsymbol{\Sigma}_z^{(r)}\right) dz_2} = \frac{\frac{\gamma_z e^{-g_z}}{2\pi^{(K-1)}|\boldsymbol{\Sigma}_z|}(K-1)!\gamma_z^{K-1}\mathscr{L}_{K-1}\left(-\frac{\|\eta\|^2}{\gamma_z}\right)}{\frac{\gamma_z e^{-g_z}}{2\pi^{(K-2)}|\boldsymbol{\Sigma}_z|}(K-2)!\gamma_z^{K-2}\mathscr{L}_{K-2}\left(-\frac{\|\eta\|^2}{\gamma_z}\right)}$$

$$= \gamma_z (K-1)\left(\frac{\mathscr{L}_{K-1}\left(-\|\eta\|^2/\gamma_z\right)}{\mathscr{L}_{K-2}\left(-\|\eta\|^2/\gamma_z\right)}\right).$$

Reference

Gradshteyn I, Ryzhik I (1980) Table of integrals, series, and products. Academic, New York

Index

C. Brombin et al., *Parametric and Nonparametric Inference for Statistical Dynamic Shape Analysis with Applications*, SpringerBriefs in Statistics,
DOI 10.1007/978-3-319-26311-3